THE LEGACY OF HANS FREUDENTHAL

THE LEGACY OF HANS FREUDENTHAL

Edited by

LEEN STREEFLAND

Freudenthal Institute, Utrecht, The Netherlands

Reprinted from Educational Studies in Mathematics 25 (1–2), 1993

KLUWER ACADEMIC PUBLISHERS

DORDRECHT / BOSTON / LONDON

Library of Congress Cataloging-in-Publication Data

A C.I.P. Catalogue record for this book is available from the Library of Congress.

ISBN 0-7923-2653-9

Published by Kluwer Academic Publishers,
P.O. Box 17, 3300 AA Dordrecht, The Netherlands.

Kluwer Academic Publishers incorporates
the publishing programmes of
D. Reidel, Martinus Nijhoff, Dr W. Junk and MTP Press.

Sold and distributed in the U.S.A. and Canada
by Kluwer Academic Publishers,
101 Philip Drive, Norwell, MA 02061, U.S.A.

In all other countries, sold and distributed
by Kluwer Academic Publishers,
P.O. Box 322, 3300 AH Dordrecht, The Netherlands.

Printed on acid-free paper

Printed in the Netherlands

TABLE OF CONTENTS

Editorial: The Legacy of Freudenthal 1

JOSETTE ADDA / Une lumière s'est éteinte. H. Freudenthal – Homo Universalis 9

FRED GOFFREE / HF: Working on Mathematics Education 21

HENK J. M. BOS / 'The Bond with Reality is Cut' – Freudenthal on the Foundations of Geometry around 1900 51

W. T. VAN EST / Hans Freudenthal (17 September 1905–13 October 1990) 59

HANS FREUDENTHAL / Thoughts on Teaching Mechanics. Didactical Phenomenology of the Concept of Force 71

A. TREFFERS / Wiskobas and Freudenthal: Realistic Mathematics Education 89

L. STREEFLAND / The Design of a Mathematics Course. A Theoretical Reflection 109

J. DE LANGE / Between End and Beginning. Mathematics Education for 12–16 Year Olds: 1987–2002 137

Book Review

Hans Freudenthal, *Revisting Mathematics Education. China Lecturers*, Mathematics Education Library, Volume 9, 1991 (CHRISTINE KEITEL) 161

Rarely, if ever, do researchers
ask, let alone investigate, whether and
to which degree 'errors' are due to education,
and what educational developers and teachers can
learn from them, although only tackling these
questions could justify this kind of research as
properly educational.

Hans Freudenthal

EDITORIAL: THE LEGACY OF FREUDENTHAL

Editing a special issue as an *homage* to Hans Freudenthal necessarily implies doing him an injustice. An injustice to his work and to the many members of the (inter)national community on mathematics and education with which he was so involved. What must be included and what left aside? Should his work on mathematics and its history be excluded because of the nature of 'Educational Studies'?

Who should beyond doubt be invited to contribute: the mondial upper ten of maths education, famous mathematicians, his followers or scholars, or who else?

Should the issue focus on the adventures of maths education in The Netherlands where Freudenthal lived for such a long time and to which he contributed so much? Or should one aim for a world wide history as a context in which to portray him?

What happens more often than not is that the preface is written at the end. This is also the case here, not to carry out one more 'antididactical inversion' as Freudenthal would have called it, but simply because there was no other choice.

Not until the end did it become clear what blueprint had emerged.

The only thing that was obvious in advance was that justice needed to be done to Hans Freudenthal's life and work. This was the leitmotiv that pevaded the preparation of this volume, albeit within the confines of an issue like the current one.

For this reason the edition contains articles on either Freudenthal's mathematical, historical and educational work, most written by Dutch scholars of his and also one by himself. There are also two contributions from international colleagues and friends.

Freudenthal's personality pervades each of the contributions, albeit to varying degrees.

In consequence a certain overlap between the different parts of the issue could neither be avoided nor was it preferable, so as not to affect the authenticity of individual contributions.

The array of contributors is headed by Josette Adda. She portrays Freudenthal as a universal human being. She does so in French, one of the languages he himself mastered so admirably.

The light she sheds on him is reflected in multifaceted manner. And indeed, there were many sides to Freudenthal. He detested arm-chair pedagogy, and pleaded incessantly for mathematics for all. He dealt with the demands of scholarship in mathematics education in a particular way. Here he proved to be the master of paradigms, of criticism and originally as well as productivity, while

Educational Studies in Mathematics **25:** 1–7, 1993.

founding a science on mathematics education, being himself a paradigm for the international community of mathematics education in many aspects.

It is striking to observe how Freudenthal's work reflects his personality, notably the historical and educational parts. Henk Bos' discusson of Freudenthal's essay review of the 8th edition of Hilbert's 'Grundlagen der Geometry' also illustrates this.

Its results, style and reception are dealt with subsequently by the author. Firstly he appreciates the essay review as a landmark. According to Bos Freudenthal achieved a level of description beyond that of chronological reports and explanations of mathematical achievements. Indeed he composed an accurate historical picture of the foundations of geometry of that time. Part of Bos' title 'The bond with reality is cut' reflects this. Moreover, this title foreshadows an important phenomenon from Freudenthal's later theoretical reflections on the constitution of mental objects as a part of learning mathematics, that is anonthologisation.

Secondly Freudenthal is set right by his scholar as far as the applied criteria for historical judgement and appreciation are concerned. It is not only mathematical excellence that counts to do justice to its historical development. A fine specimen indeed of the scholar who teaches the master.

Van Est takes his place among the array of contributors to this volume on Freudenthal's mathematical work. Indeed his portrait would be incomplete without it.

"There are scientists and scholars", is the first brush stroke Van Est makes and he immediately adds: "Hans Freudenthal was a scholar". Indeed he was, but not only in the sense Van Est means, as one whose learning is extensive and exact, or whose approach to learning is critical and scrupulous. If that were so this editor would have a word-usage problem. The scholar Freudenthal was also a pupil. And, it is not the worst characteristic of a scholar to also be one who learns.

Van Est gives a concise image of Freudenthal's life. Mathematics not only runs through it like a continuous thread, but also acts for the man we commemorate as the dominant key to synthesize an extensive range of knowledge. This is how the author characterizes one of the most powerful of Freudenthal's abilities.

His work as a lecturer on many mathematical topics passes the review, first his period in Amsterdam and later on in Utrecht, with World War II in between.

In an appendix Van Elst illuminates Freudenthal's work on Homotopy Theory and Lie Groups in Geometry. Although this is daily fare for insiders it must be said that the author's description makes it accessible for a broader readership. With respect to the subjects mentioned he shows how Freudenthal contributed substantially to their development.

Topology marked the beginning of Freudenthal's scientific career. The title of his thesis 'Über die Enden topologischer Räume und Gruppen' reflects his contribution to it. Later 'these ends' became known among professionals as 'Freudenthal-ends'.

Indeed, starting with ends seems to be more than a coincidence, because it was something Freudenthal did.

It was only a week before he passed away that Freudenthal put the finishing touches to 'Revisiting Mathematics Education. China Lectures', the fourth and final part of his opus magnum on the subject. Alan Bishop prefaced it as the 'definite' Freudenthal. Its motive evolved from a journey to China he made at the age of 82. There he encountered several constraints in the community of mathematics education to understand his special approach to it. That is why he polished up his educational heritage as a beacon for his followers and scholars, even if they would come from China. According to him it added nothing but itself to what he wrote before.

Christine Keitel discussed the book. Her discourse is both lucid and concise. She illuminates the mathematical-didactical working of his mind. But in doing so she hits upon some problems in his work. One of them concerns the tension between the demand for structuring tools to organise the field and Freudenthal's deliberate refusal to accept to structure it in a systematic manner. Another relates to the vagueness and elusiveness of concepts such as 'common sense', which according to Freudenthal's course of thought, is not only the driving force of learning mathematics, but is itself developed by doing mathematics.

It will be the future task of the institute named after him to solve these problems, according to the reviewer.

Freudenthal's book is not the only milestone he himself erected for the post-Freudenthal era. Another smaller part of his recent work exists which remained unpbulished thus far internationally. Again its motive is very much Freudenthalian.

What started as an attempt to review a dissertation on the changes of the teaching of mechanics in secondary education in The Netherlands expanded into a didactical phenomenology of the concept of force. There is no need to guess at how he analyses the phenomenon under discussion. Indeed, it notably reflects his approach to mathematics education. Therefore it is about *mechanisming* rather than *learning mechanics.* It contains original views such as on rest being a limit case of motion rather than its opposite. Moreover pupil's experiences from daily life should not be suppressed, even when these experiences interfere rather inconveniently with the scientific concepts aimed at in physics teaching.

On the contrary, these experiences should be recognised as such and be exploited as starting points for learning processes.

According to the author it is necessary that the learner itself – albeit under the teacher's guidance – reforms his intuitive experiences into more scientific ones. That is why Freudenthal's suggestions for activities appeal to and start with those experiences.

Gradually the focus of attention is shifted from subjective forces to more objective ones, from the static to the kinematic aspects of force and also from the objects themselves to somewhere in between, while interpreting forces as sides of single interactions and refraining from seeing them as actions and reactions.

Fred Goffree peruses Freudenthal's work on mathematics educaton and this was no easy job. The author consulted many sources to sketch H.F., even some yet unpublished. Indeed, H.F. he calls him, after H.F. himself. Of course, it is very difficult, if not impossible, to provide for a chapter that does not reflect the same H.F. as the one who pervades all the other chapters in this monograph.

Offering an impression of H.F.'s working on mathematics education means emphasising the final part of his professional life and even beyond. But he indeed also had his prehistory. Lessons given to his two young sons made him reflect on arithmetic at primary level. This gave motive to the writing of his didactics of arithmetic in 1942. Several of his later important ideas are rooted in this unpublished document.

Many interesting episodes in H.F.'s life pass the review, too many to summarize. Let us just take a few snapshots.

After World War II, H.F. became a member of the Mathematics Study Group. For this group he once lectured on 'Education in thinking'. It fascinated him to think about the thought process, or so it seemed. At first his perspective was more theoretical than practical. The shift was ushered in by his becoming stuck in the dilemma of whether it was he going around in circles or the world he wanted to observe!

On other occasions he was sailing his own steady course rather than circling around. This, for instance, was the case when he discussed geometry education with Tatjana Ehrenfest. He already then – in 1951 – understood the formal value of geometry qualifying geometrical orientation in the world around us – which means organising it with simple geometrical means for better understanding – as valuable in itself.

Sometimes others determined the course to sail or at least influenced it to a certain extent. This was the case with the Van Hieles and their work. His thinking on learning processes and their importance was encouraged by them and their work, especially by P. van Hiele's level theory, although H.F.'s ideas about levels were different and more subtle, as he would later demonstrate. How important learning processes were according to him is shown by how he disseminated the need of observing them, time and again, if not incessantly.

In the sixties Freudenthal took the membership of the Modernization Committee of the Mathematics Curriculum (CMLW). From then on he set to work on the development of mathematics education with increasing intensity, meanwhile propagating his scepticism about laws and royal decrees to change it. That is why he invested his efforts towards changing teachers and textbooks and to encouraging and asking others to do the same. What it led to? Well, to 'H.F.: Working on mathematics education'.

Adrian Treffers describes Freudenthal's influence on the development of realistic mathematics education at primary level in The Netherlands. This regards the Wiskobas-project of the former IOWO in particular. Freudenthal's name has been closely linked to this project. It is explained why the truth with respect to the suggested connection varies from being correct and incorrect.

Next, an intriguing problem which starts with a newspaper clipping about Van Gogh's sunflowers in a wheat-field pervades the author's discourse as a guiding example. Rich thematic contexts, long-term learning strands and educational development pass in review, each time followed by reflections on if and how Freudenthal influenced the actual subject.

Rich thematic contexts, for instance, were important according to him. He predicted the disappearance of mathematics as individual subject matter in education sometime around the year 2000, because it would by then have merged with integrated thematic contexts.

This point of view, elaborated in several publications, self-evidently had its influence on the work of Wiskobas.

As far as long term learning strands on basic skills and column arithmetic are concerned, Freudenthal shifts from actor to observer. In this respect he in turn was influenced by Wiskobas. In first instance he hardly had an eye for the importance of memorised addition and multiplication tables. The author observes that Freudenthal devoted not a single thought to the change of perspective from column arithmetic to the merit of flexible mental computation in his last book, as was urged by Wiskobas, or so it seems. It must be said that Freudenthal, more often than not, approached the memorization of factual, mathematical knowledge and the training of skills in an indirect manner.

Indeed he kept on pleading to keep open the sources of insight for the learner in the teaching-learning process. This was a main concern of his. Moreover for a very long time he could not put aside the conflict between training skills and the retention of insight. Finally a solution seemed to emerge. As far as fractions and ratio are concerned, Freudenthal's influence was considerable. His didactical phenomenologies were landmarks for how and where to seek in order to outline the subjects realistically but at the same time his expositions were labyrinthine for designers of educational material because of the overwhelming number of ideas he offered them. Clear outlines for learning strands remained concealed. Developments on measurement and geometry were influenced decisively by him.

Finally, Treffers addresses the question of educational development and developmental research. Many aspects related to Freudenthal's personality remain underexposed, but by giving a summary of them his portrait nears completion.

On 29 March 1985 Hans Freudenthal received the honorary membership of the NVORWO (Dutch Association for the Development of Primary Mathematics Education).

On this occasion his words of thanks to the annual general meeting were tape-recorded due to a stay abroad. He addressed the matter of geometry, which was the theme of the day. In referring to his publications he said:

> Actually I should – if time permitted – write a brand-new chapter on geometry. What I would then emphasize, is what Streefland calls the intertwining of learning strands, in this instance – roughly spoken – the arithmetical and the geometrical; an early intertwinement for the sake of the anticipatory learning of concepts and actions that will develop in full at a later time.

Later Freudenthal re-emphasized the importance of my idea of intertwining or interweaving learning strands more than once. In consequence this editor was both inspired and encouraged to collect his threads of thought on the topic in the early eighties, albeit within the broader theoretical framework of 'The design of a mathematics course'. There the intertwinement of learning strands plays a considerable role, as does change of perspective, because of the shifting from one learning strand to another and vice versa, while keeping an eye on the goals of the teaching-leanring process one has in mind to pursue.

These, among others, are the strategies to be applied by the designer of a course who, moreover, later wishes to intertwine the strands of both teaching and learning on which he anticipates. In his final book Freudenthal accuses himself of not having paid sufficient attention to change of perspective. Hopefully it is in his spirit that this has been addressed here.

In the year 2000 there will be no more mathematics education. Textbooks for mathematics at school level will have disappeared and the subject will have vanished completely.

This provocative prediction Freudenthal made on the occasion of his retirement as IOWO's director. I referred to it earlier in connection with Treffers' contribution.

It was not gloom that prompted this prognosis. On the contrary. Freudenthal did neither predict nor plead for the disappearance of mathematics in schools.

He only foresaw its integration in other subjects, its seamless transition into other fields. Mathematics, as Freudenthal stated in 1976, exists 'to be gone through and to be indulged, like reading, writing, pottering, drawing, singing and breathing, that is in integrated education'.

Sixteen years later it turns out that this dream will not come true. After the succession of centuries mathematics will not be wiped out from schedules, as Jan de Lange shows in his contribution.

But it is argued by the author that the character of mathematics education will have changed notably by then as current changes foreshadow.

He looks ahead to mathematics in the year 2002.

Examples and experiences from two projects of the Freudenthal Institute act, among several others, as vehicles of thought. The author's discourse and annotations on Freudenthal's ideas are alternating.

To extract a framework of reference 'to put Freudenthal in' was not at all a difficult job. On the contrary, Freudenthal would very much have appreciated the examples, especially the ones on geometry, either with a retrospective or a perspective function. Both examples are connected with recent developments, difficulties that teachers encounter while shifting from traditional to realistic mathematics education.

Freudenthal follows as it were the course of thought from the margin. On several occasions he intervenes, albeit by means of his publicatons and the ideas he presented in them.

In his contribution Goffree refers to Freudenthal's 'Werken aan onderwijs: op weg' (Working on education: underway). It is there that Freudenthal ends with:

> Yes, I was a drummer, but to deaf ears.
>
> This reminds me of a poem by Heine (disappointing that I have to finish in German).

Schlage die Trommel und fürchte dich nicht
und küsse die Marketenderin!
Das ist die ganze Wissenschaft,
das ist der Bücher tiefster Sinn.

Trommele die Leute aus dem Schlaf,
trommele Reveille und Jugendkraft,
marschiere trommelnd immer voran!
Das ist die ganze Wissenschaft.

Das ist die Hegelsche Philosophie,
das ist der Bücher tiefster Sinn!
Ich hab' es begriffen, weil ich gescheit,
weil ich ein guter Tambour bin.

The poem speaks of a drummer who drums tenaciously. He keeps bringing forth his penetrating sound despite people, books, science, philosophy. Somehow it reflects Freudenthal's way of working: achieving notable results with simple means.

Then Freudenthal procees with:

> No, I was not a good drummer. A good drummer does not drum at night, but in the morning when it is time to get up. But don't think I have regrets. An enfant terrible should be proud of his vices.

Was Freudenthal not being too modest here? He did indeed beat the drum for mathematics and math education. But did he do so at night, in the dark? Or must we blame outselves for being blind to the abudnance of light he gave us in this respect? Not all of us! Fortunately many from the community of mathematics education came forward as his treasurers. Some witnessed of their stewardship in this special issue. In doing so they presented their 'Legacy of Hans Freudenthal'.

ACKNOWLEDGEMENTS

This issue could be achieved thanks to the cooperation of the authors, the Freudenthal family for giving their permission for the posthumous publications of Freudenthal's article on force, the Freudenthal institute for the advice of colleagues', the editorial support by Mrs. Marianne Moonen and finally translator Mrs. Hetty Vonk for her support and advice. I wish to express my gratitude to all of them for enabling me to coordinate this commemoration of Freudenthal.

Freudenthal Institute
Utrecht University
Tiberdreef 4, 3561 GG Utrecht
The Netherlands

LEEN STREEFLAND

JOSETTE ADDA

UNE LUMIÈRE S'EST ÉTEINTE
H. FREUDENTHAL – HOMO UNIVERSALIS

ABSTRACT. We intend to present here an homage to the humanistic aspect of Hans Freudenthal's personality. His culture was devoid of any frontier and he always struggled (in many languages) against obscurantism. His thoughts and his works went in many complementary directions: mathematics, history of mathematics, mathematics education, philosophy... He worked to open mathematics education to everyone and never lost the intellectual requirements of a great scientific thinker. But he was also a man of action and had a great influence on the development of mathematics education research, not only in The Netherlands but also all around the world.

RÉSUMÉ. Nous voulons rendre ici hommage plus particulièrement à l'aspect humaniste de la personnalité de Hans Freudenthal. C'était un homme dont l'immense culture ne reconnaissait pas les frontières. Il s'est opposé (et ce dans plusieurs langues) à tous les obscurantismes. Son œuvre s'est développée dans un très grand nombre de directions: mathématiques, histoire des sciences, éducation, philosophie, ... Il a œuvré pour l'ouverture de l'enseignement des mathématiques 'pour tous' et pourtant il a toujours suivi les exigences intellectuelles d'un grand homme de sciences. Il fut aussi un homme d'action et a joué un rôle essentiel dans le développement de la recherche sur l'enseignement des mathématiques non seulement aux Pays-Bas mais dans le monde entier.

H. F. – UN HOMME SANS FRONTIÈRES

Qu'il me soit permis de rendre hommage ici[1] à H. Freudenthal en français: il maîtrisait admirablement cette langue comme bien d'autres: néerlandais, allemand, russe, anglais, hébreu, j'en oublie probablement et je crois bien qu'il avait commencé à apprendre le chinois pendant le long voyage en Chine qu'il a effectué à 83 ans! Il était capable de tenir une conférence dans plusieurs langues simultanément et reconnaissait simplement, les dernières années, qu'à partir de trois langues simultanées, il commençait à avoir quelques troubles en descendant des escaliers. Surtout chacun admirait les finesses de son érudition linguistique: son français était riche de sa grande connaissance du latin et de notre littérature de toutes les époques. En fait pour lui, il n'y avait pas, je pense, une littérature française, une littérature allemande, une littérature anglaise, etc., sa culture immense réalisait une véritable synthèse des littératures, des peintures, des architectures, etc... de toutes les nations. La dernière fois que je l'ai vu, à Paris, en septembre 1989, après avoir fait une conférence d'histoire des mathématiques il protestait avec passion contre le violent éclairage des tableaux au Musée d'Orsay et s'apprêtait à écrire à son directeur!

J'ai repris en titre l'expression "Homo Universalis" dont l'a qualifié le président de l'Université d'Utrecht lors des cérémonies d'ouverture de l'Institut Freudenthal;

Educational Studies in Mathematics **25:** 9–19, 1993.

je trouve qu'on ne peut rendre mieux sa principale caractéristique: un homme véritablement sans frontières, ouvert à toutes les cultures, aimant citer Joyce, Goethe ou Platon; tolérant à tous les points de vue philosophiques, c'était un authentique descendant de l'Esprit des Lumières et il lutta toute sa vie contre tous les obscurantismes.

Hans FREUDENTHAL naquit le 17 septembre 1905 à Luckenwalde en Allemagne. Il prit sa retraite de l'Université d'Utrecht en 1975; le 'feestboek' de son jubilé (septembre 1975) comportait une liste de publications de 380 titres avec une grande variété de langues (sans parler bien sûr des traductions). Mais il n'a pas arrêté de travailler à ce moment, il a continué à réfléchir et à écrire. Il est décédé le 13 octobre 1990, une semaine après avoir relu les épreuves de son dernier livre paru en septembre 1991 (éditions Kluwer) "*Revisiting Mathematics Education–China Lectures*".

H. F. – UNE ŒUVRE MULTIDIMENSIONNELLE

Il avait soutenu sa thèse de doctorat en mathématiques à l'Université de Berlin en 1930 ('*Über die Enden topologischer Raüme und Gruppen*') et son habilitation au titre de 'privaatdocent' à l'Université d'Amsterdam en 1931 ('*Qualität und Quantität in der Mathematik*').

Il fut assistant de Brouwer et, bien plus tard, il fera sur ses travaux des recherches historiques qu'il nous racontera lors d'une conférence à Paris[2]:

> J'avais presque achevé le manuscrit de l'œuvre topologique de Brouwer quand, à la recherche d'échantillons de l'écriture de Brouwer qui se prêteraient à la reproduction, il me survint qu'un cahier tombant à la terre lâcha quelques pages détachées que j'identifiai immédiatement comme des brouillons de lettres à Hilbert et Hadamard.
>
> D'après ce brouillon, Brouwer passait les vacances de Noël de 1909–10 chez son frère, le géologue, 6, rue de l'Abbé de l'Epée, au quartier latin de Paris. Voilà donc le berceau de la topologie moderne dont la naissance fut annoncée au Nouvel An 1910. En effet, ces brouillons contiennent la majorité de ce qui est essentiel dans l'œuvre de Brouwer telle qu'elle sera publiée en 1911-13, ...

D'autres articles rendront compte de son œuvre mathématique et de son œuvre didactique; je voudrais seulement montrer ici, par quelques examples de titres, l'immense variété des thèmes traités: '*La structure des groupes à deux bouts et des groupes triplement transitifs*' (1951), '*Gambling with a poor chance of gain*' (1952), '*La topologie dans les fondements de la géométrie*', Proc. Int. Congress Math 1954 et les livres '*Logique mathématique appliquée*', Gauthier-Villars 1958, '*The language of logic*', Elsevier 1966, '*Probability and statistics*', Elsevier 1965 mais aussi les articles '*Is there a specific problem of application for probability*' in Mind 1941, '*Machines pensantes*', conférence du Palais de la Découverte 1953, '*Relations entre l'enseignement secondaire et l'enseignement universitaire en Hollande*' in Enseignement mathématique 1956, '*Ist die mathematische Statistik Paradox?*' in Dialectica 12 – 1958, '*Sur la responsabilié humaine du mathématicien*' in Mathematica & Paedagogie 1957–58, '*Abus philosophiques de la statistique*', in Revue de Métaphysique et de Morale – 1962, '*Les faits et gestes de l'âne de Buridan*', XII Congrès International d'Histoire des Sciences Paris 1968, '*Notations mathématiques*' in Encyclopaedia Universalis 1972 et le livre si étonnant, malheureusement épuisé: '*LINCO, design of a language for cosmic intercourse*', Amsterdam 1960 où cet Homo Universalis allait jusqu'à imaginer de communiquer avec des êtres d'autres planètes.

Comme cet autre grand humaniste qu'avait été A. Einstein à qui il ressemblait d'ailleurs de manière frappante pas seulement par sa chevelure frisée mais aussi par son regard pétillant, il était toujours à l'écoute de tous ses semblables et attentif à tous les problèmes humains et sociaux; c'est ainsi que très naturellement il se mit à s'intéresser de plus en plus aux problèmes éducatifs. Il m'a souvent dit qu'il avait déjà cette passion lorsqu'il était jeune, assistant de mathématiques et qu'il avait organisé à cette époque un séminaire sur l'enseignement.

Il fut membre du Comité exécutif de la Commission Internationale pour l'Enseignement des Mathématiques (ICMI) de 1963 à 1974 et son président de 1967 à 1970. Il fut l'un des plus actifs participants de toutes les rencontres internationales sur l'enseignement des mathématiques (notamment de la CIEAEM dont il fut aussi président et de PME qu'il contribua à fonder). Il écrivit énormément d'articles sur ces sujets et ces livres d'une importance exceptionnelle: '*Mathematics as an educational task*' (1973), '*Weeding and Sowing – Preface to a Science of Mathematical Education*' (1978), '*Didactical phenomenology of mathematical structures*' (1983) et son œuvre posthume '*Revisiting mathematics education*' (1991) qui reprend, développe, explicite et synthétise son message comme un testament spirituel.

Mais son œuvre n'est pas seulement constituée de pages écrites et d'idées, c'est aussi (car il était également un homme d'action) la création de IOWO à Utrecht, qu'il vécut vraiment comme un 'enfantement', dans lequel ils'engagea avec passion, qu'il craignit plusieurs fois de voir disparaître mais qui avait perduré sous le nom de OW&OC; il était juste que cet Institut porte enfin depuis 1991 le nom par lequel le monde entier le désignait déjà: 'Institut Freudenthal'. Au plan international, il a fondé Educational Studies in Mathematics en 1968, s'est dépensé sans compter pour son développement et, une fois l'audience et la valeur scientifique assurées, il a tenu à s'effacer et a passé le relais de la direction du Comité de rédaction à A. Bishop en 1979.

C'est lui qui a poussé à la création de PME (en 1976) en organisant son premier Congrès à Utrecht dans un esprit interdisciplinaire pour une meilleure connaissance réciproque entre psychologues et mathématiciens.

H. F. ET LES 'MATHÉMATIQUES POUR TOUS'

Il n'appréciait pas du tout ce qu'il appelait 'arm-chair pedagogy', la didactique en fauteuil ou en bibliothèque, pour lui et pour ses disciples il n'y a pas de didactique sans élèves: il aimait à dire qu'il apprenait énormément sur les mathématiques lorsqu'il se trouvait dans une classe élémentaire. Lorsqu'il fut invité par le Congrès de l'Association des Professeurs de Mathématiques de l'Enseignement Public à Lille (1983) sur 'l'échec en mathématiques', il nous fit une merveilleuse conférence[3] dont je veux extraire quelques citations:

> *L'échec des coureurs*
> L'échec, c'est relatif. Si l'on se poste pour une course et l'on n'est pas même classé, c'est un échec. C'est l'échec des coureurs à pied, à bicyclette, à cheval, à automobile, des patineurs, des boxeurs. Mais personne ne vous demande à courir, même à une vitesse modérée, à monter à bicyclette, à cheval, en automobile, à patiner, à boxer, même pas à jouer aux échecs où les échecs peuvent être bien spécifiques. Qu'ils se plaignent de leurs échecs, cela ne me touche guère. C'est bien leur faute à eux s'ils échouent. Les mathématiques, c'est autre chose. Il n'y a pas de choix. Apprendre les mathématiques, apprendre à lire et écrire c'est obligatoire...
>
> Voilà ce que j'ai l'habitude d'appeler le gros mensonge de notre système d'enseignement – et j'ai peur qu'il ne caractérise bien d'autres systèmes: l'abîme qui s'ouvre entre les demandes exagérées et le niveau bas de performance qui suffit pour être admis. Un gros mensonge, qui n'est qu'un des aspects des fonctions sociales de notre système d'enseignement – je parle de chez nous en tant que de paradigme. Des examens sont un outil de sélection: d'identifier ceux qui peuvent satisfaire aux demandes les plus excessives? Mais pourquoi la grande majorité serait-elle obligée de tendre à infiniment plus haut qu'ils peuvent atteindre? A mon avis apprendre des mathématiques simples à un niveau élevé d'entendement vaut mieux qu'enseigner des mathématiques raffinées qui s'apprennent à un niveau bas au dessous de chaque niveau possible.
>
> Je n'aspire pas à changer la société, mais malgré des désillusions que j'ai éprouvées, je crois toujours qu'on peut changer peu à peu l'enseignement: au lieu de course avec l'échec inévitable des coureurs, la promenade où la modération de la marche est récompensée par l'attention pour l'entourage et ses détails qui échappe à celle des coureurs – excusez cette abondance de métaphores, mais n'oubliez pas que le langage métaphorique a déjà commencé avec "la lutte contre l'échec".
> Les promenades en groupe ont l'avantage qu'on s'amuse en apprenant l'un de l'autre. J'ai fait une telle promenade pendant presque trois années. Ma compagne était une fille de près de chez nous. Je lui ai donné des leçons, à une moyenne de deux fois par semaine une demi-heure ou un petit peu de plus. Ce fut dans sa sixième année scolaire, la dernière de notre enseignement primaire, que cela commença. Elle était d'une intelligence normale, mais extrêmement faible en arithmétique à l'école.

Ce qu'il ne dit pas c'est qu'il s'agissait d'une fillette extrêmement défavorisée socialement et immigrée depuis peu du Surinam aux Pays-Bas. Il nous 'montra' (brouillons à l'appui) chez cette enfant une évolution que l'on pourrait trouver fascinante et pourtant il garda la modestie et la lucidité du scientifique:

> Je vous ai avertis: la leçon de cette suite de leçons n'est pas grand-chose. C'est une source de questions plutôt que de réponses, mais puis – nous l'espérons – de questions qui valent la peine d'être posées. Des questions de méthode et de matière à enseigner. De méthode – c'est évident: l'organisation, si elle est possible, d'un enseignement qui approche celui que j'ai eu la bonne chance de donner à cette fille. Mais n'oublions pas la matière à enseigner. Quel est le sens de cet enseignement de fractions, de nombres négatifs, d'équations linéaires, de puissances, d'algèbre, pour une fille qui ne sera jamais appelée à appliquer cette matière? On s'est amusés, le maître et l'élève. Peut-être elle a gagné plus de confiance en elle-même. Ce seraient des résultats positifs, mais cela ne suffit pas. C'est évident que tout le monde ne peut pas apprendre les mêmes choses, mais l'algèbre, doit-elle être abordée par tout le monde, et puis, quelle espèce d'algèbre, quelle quantité, quelle profondeur? Evidemment l'algèbre ne vaut pas une course, mais vaut-elle promenade?
> Si je n'aime pas l'enseignement en pleine classe dirigé par le maître, je me méfie de l'enseignement dit individualisé qui, nonobstant des résultats douteux, est toujours à la mode – au moins chez nous – et dont la menace sera renforcée au futur par l'intervention de l'ordinateur. C'est la promenade à seul transformée dans une course, qui produit de nouveau des foules de décrocheurs. J'aime plutôt les petits groupes – même et par préférence hétérogènes – de travail, la différentation non par rapport à la matière à enseigner et à la vitesse d'apprentissage, mais par rapport à la profondeur de l'entendement des participants....

Il resta toujours ferme sur cette position en faveur des classes hétérogènes et poussa au développement des recherches didactiques sur l'organisation de groupes de travail hétérogènes. Son épouse était d'ailleurs une fervente militante du mouvement pédagogique dit de 'l'école d'Iena', école dont les classes mêlent des enfants d'âges différents comme dans les familles.

Il est regrettable que l'on ne possède pas de traduction en français ou en anglais du récit de ses '*promenades avec Bastien*' où il rapporta, jour après jour, l'aventure intellectuelle de ses promenades avec son petit-fils, bébé et jusqu'à sa scolarisation. Il en citait souvent des anecdotes qui lui ont servi comme contre-exemples à la théorie de Piaget en mettant en lumière les potentialités réelles des enfants en situations naturelles (i.e. non 'expérimentales' comme celles de Piaget et son école). Tous les didacticiens ont en mémoire d'innombrables aventures mathématiques à la fois fascinantes et drôles qu'il nous racontait au cours de conférences mais aussi de repas, de promenades, ...

H. F. ET L'EXIGENCE SCIENTIFIQUE

Je sais bien que certains didacticiens d'une tendance que je qualifierai de 'scientiste' lui ont fait grief de cette utilisation d'anecdotes; c'est là ne pas avoir compris du tout sa pensée: dès '*Mathematics as an educational task*' il expliquait très bien le rôle scientifique des 'exemples paradigmatiques', or toutes ces anecdotes en étaient et permettaient de présenter, de manière simple, accessible à tous, sans jamais le moindre discours jargonneux, des *phénomènes* importants. C'est ailleurs lui, en fait, qui a donné leur essor aux études phénoménologiques en didactique des mathématiques. Par dessus tout, il était très attentif à l'importance du *contexte* (son influence positive ou négative dans l'apprentissage ainsi que les artefacts des expériences, ce qui le rendait très sévère vis-à-vis de multiples prétendus travaux scientifiques) au point que ses disciples avaient composé pour son jubilé une chanson dont le refrain '*Catch the context*' l'a rempli de joie.

Dans son allocation au 1er Congrès International de l'Enseignement mathématique à Lyon en 1969, il soulignait déjà l'opposition qui lui sera chère entre '*mathématiques toutes faites*' et '*mathématiques à faire*':

> Les mathématiques sont plus qu'une technique. Apprendre les mathématiques, c'est acquérir une attitude de comportement mathématique. Les mathématiques n'ont pas être enseignées pour s'adapter à une minorité mais pour tous et ils doivent apprendre non seulement des mathématiques mais aussi quoi faire avec des mathématiques. Ceci ne signifie pas enseigner des mathématiques appliquées mais plutôt créer une attitude, l'attitude de découvrir des mathématiques partout où elles se trouvent.

Aux cérémonies de son jubilé (août 1976), interrogé sur ce que serait, selon lui, l'enseignement des mathématiques en l'an 2000, il répondait:

> La réponse est toute simple – il n'existera pas; il aura disparu. Il n'y aura pas de matière d'enseignement appelée "mathématiques", il n'y aura pas de leçons de mathématiques sur l'emploi du temps et il n'y aura pas de manuels de mathématiques.

Il essayait de promouvoir dans les recherches de IOWO, puis de OW&OC, un enseignement intégré, une formation transdisciplinaire mais il ne voulut jamais abuser de son autorité pour forcer l'évolution de ces institutions. Il me parla longuement déjà en 1971 de son horreur de la 'gérontocratie' et de son désir de laisser ses jeunes collègues travailler librement (éventuellement en faisant des erreurs), il expliquait qu'il préférait être le 'grand-père' plutôt que le 'père', et ne donner des conseils que si l'on insistait pour lui en demander.

Voilà l'un des traits de caractère de H. Freudenthal qu'il est essentiel d'évoquer pour comprendre sa personnalité et son œuvre: son sens des responsabilités. Dans '*Weeding and Sowing*', il insiste sur la différence entre 'accountability' et 'accountancy'.

G. Howson dira, lors de la cérémonie pour les 80 ans de H. Freudenthal (septembre 1985):

> ... Freudenthal has rarely opted for safety and appeasement. Indeed it is this which brings me to the third point which I wish to mention and where I believe Freudenthal has made a unique contribution to mathematics education; that is the time he has devoted to the criticism of the work of others.
>
> A remarkable feature of the 1960s and 1970s was the number of mathematics educators who produced a variety of schemes, ideas, texts, etc. and who

> did so without any apparent recognition that others were doing so too and that there were remarkable differences in assumptions, goals and attempted solutions. In general there was an air of 'dog does not eat dog' – one educator did not criticise another, and everyone went on with his own work largely oblivious of the others. This was not an atmosphere conducive to laying the foundations of a discipline. One definition of 'discipline' given in my dictionary is 'training through correction and suffering'. Often in the 1960s and 1970s it was the children who suffered because of the lack of correction so far as the mathematics educators were concerned. Serious, detailed criticism is not easily carried out, and it takes one away from those activities which more readily gain general esteem. Freudenthals' biting criticism of the way in which much empirical research work in mathematics education is carried out and, more importantly, reported and quoted, must have taken hours and hours of work which in itself carried little personal reward or satisfaction. Yet that work had to be done in order to assemble a strong case against the system – a case presented by means of an argument which justified the word 'discipline'. Another example is the 1975 paper Pupils Achievements Internationally Compared, a sixty page critique of the First International Study on Mathematical Attainment. Such critical work is, as I remarked earlier, unfashionable and certainly it is not guaranteed to make one many new friends, but wihtout it there is no chance that we shall ever attain to that science of mathematics education which Freudenthal's '*Weeding and Sowing*' sought to preface.

Il osa s'attaquer aux deux grands du moment (Bourbaki et Piaget) par cet exergue:

> His mathematics stemmed from Piaget, and his developmental psychology from Bourbaki. He did not like complete induction so he preferred complete intimidation.
>
> From the biography of a didactician

à l'un des chapitres de '*Mathematics as an educational task*', dont l'appendice mettait en pièces '*Piaget and the Piaget schools investigations on the development of mathematical notions*'.

Il dénonça les abus de l'utilisation des statistiques:

> According to a well-worn joke there are three kinds of lies: lies, damned lies, and statistics. When the joke was invented, mathematical statistics had not yet come into being. Otherwise one would have added three kinds of surreptitious credibility: by word of honour, by oath, and by mathematical statistics. Or three methods to assume a learned air: footnotes, bibliography, and correlation tables.

Il fut vraiment féroce avec la 1ère enquête de l'IEA et se fit aussi beaucoup d'ennemis en s'attaquant à la taxonomie de Bloom:

> How could all this develop and never be resisted? How to explain the tremendous success of the 'Taxonomy'? How the fashion of instructonal objectives, the rage of atomization, the cult of packaging, the rituals? How could educa-

> tionists who have nothing to offer but empty boxes and slogans get settled in the educational system, first in the United States, but in an increasing density in Europe too? Empty vessels make the most noise according to a proverb.

à la pédagogie par objectifs et aux modes ('mastery learning', etc.). Il s'attacha à faire connaître les travaux de Kamin ('The science and politics of IQ.') dénonçant ainsi les publications de Burt sur les soi-disant proportions (80/20!) d'inné et d'acquis qui révulsaient son humanisme.

Vis-à-vis de la Réforme des mathématiques-modernes (cf. son article dès 1963: '*Enseignement des Mathématiques modernes ou Enseignement moderne des Mathématiques*' in L'enseignement mathématique), il garda toujours le regret de ne pas s'être rendu au fameux Colloque de Royaumont en 1958. Sa position était très critique: il nous la rappela dans son débat avec Madame Krygowska sur 'la réforme de l'enseignement des mathématiques' organisé par la Régionale Parisienne de l'Association des Professeurs de l'Enseignement Public en Avril 1975[5].

> Quelqu'un a parlé des erreurs. On peut le faire dans diverses perspectives. Il y a les petites erreurs des manuels, mais on peut aussi faire une réforme qui soit globalement une erreur. Et il ne faut pas objecter la réforme continue, parce qu'on peut continuer les erreurs! (rires)
> Faire une réforme est une grande responsabilité, il faut la préparer sérieusement. On a parlé de points de vue différents, mais il y a aussi des points de vue faux, comme celui, en pédagogie, de partir de l'abstraction pour aller aux fausses concrétisations.
> C'est aussi une erreur de penser qu'on peut faire une réforme sans préparer les maîtres. Cette erreur a été faite en France et en Allemagne. Il faut une stratégie pour préparer les gens, maîtres et parents. Il faut employer des moyens modernes, comme la télévision...

D'ailleurs, à la Rencontre CIEAEM de Sherbrooke (1987) sur 'le rôle de l'erreur dans l'apprentissage et l'enseignement de la mathématique', il choisit de parler de '*Erreurs du professeur – Analyse didactique de soi-même*'[4] et déclara notamment:

> Je prends mon début avec une erreur historiquement notable d'apprentissage mathématique, l'erreur du siècle, pour ainsi dire, qui, tout en faillant ruiner profondément l'enseignement mathématique, a néanmoins réussi à ébranler la croyance qu'on y mettait et à retarder son renouveau urgent. Vous comprendrez ce que j'ai en vue: la mathématique bourbakiste parachutée à l'école, Royaumont 1959. Mais Royaumont n'en était pas le berceau. On peut suivre les traces de sa préhistoire et de ses conséquences même dans ces entretiens de notre CIEAEM.
> Cette erreur était – elle une erreur d'apprentissage? Oui, une erreur d'apprentissage mathématique au sens que, pour la plupart des participants à de nombreuses réunions, cette mathématique fut une nouveauté mal comprise. Mais surtout ce fut une erreur didactique du professeur–élève se soumettant aux ordres du professeur–professeur.
> Même à cette époque-là, chaque didacticien des mathématiques ou même

> chaque mathématicien savait ou devait savoir que la déductivité, loin d'être le point de départ, n'est que la fin élaborée de chaque activité mathématique, mais jamais, avant ou après, la didactique des mathématiques n'a été trahie en faveur d'un dogmatisme aussi despotique qu'à cette époque-là. Quels étaient les coupables de ce presqu'assassinat de la didactique des mathématiques, de cette victoire remportée à Royaumont par les dogmatiques sur l'enseignement mathématique? ...

Dans '*Revisiting Mathematics Education*', il reprend ces critiques au nom du 'common sense':

> At present most people will agree that New Math's contempt for common sense has been a historical mistake. But has everybody really learned the lesson?

Au congrès PME de Montréal de 1987, il avait été effaré par l'importance prise par les discours sur la théorie constructiviste, il lui consacre dans '*Revisiting Mathematics Education*' un long paragraphe explicite dont je citerai ces quelques lignes:

> Would any scientist who lacks faith in objective knowledge have ever had Voyager 2 launched for a visit of the big planets – as far as Neptune – to gather knowledge about them? Would he have had big accelerators built that are to discover missing elementary particles? Would he analyse long DNA strings in order to manipulate them? Would he excavate ancient cities and decipher ancient inscriptions? These are just a few questions out of an interminable variety, and yet I did not even mention the mathematician searching for a new knowledge.
>
> Rather than trying to uncover the scientist's implicit philosophy and methodology, philosophers and methodologists of science, if unacquainted with real science or frustrated by it, dream up their own fancies. To be sure, Von Glasersfeld is in good company, but in a company of spectators rather than of actors in science. Yet, in his particular case, it is a third kind of philosophy that should count: philosophy of education. These three may be mutually related provided they are not confused with each other. At any rate, I cannot see any bond between mathematics instruction on the one hand and an alleged or assumed lack of faith in an objective mathematical knowledge on the other hand, whether it is called constructivism or anything else.

Il se sentait le devoir de dénoncer toutes les erreurs et restait pourtant d'un constant optimisme, puisqu'il croyait vraiment pouvoir les déraciner. Il avait une confiance peut-être trop naïve dans l'avenir et naturellement il lui est arrivé de se tromper: par exemple, dans l'article "*Notation mathématique*" publié dans l'Encyclopœdia Universalis, il y a 20 ans, il écrivait:

> L'évolution qui a conduit à ce style utilisant les quantificateurs s'est faite peu à peu. Elle est loin d'être achevée. Prenons un exemple qui est encore reçu, mais qui ne le sera plus dans vingt ans. Pour démontrer que $\sqrt{2}$ est irrationnel, on raisonnerait ainsi: "Supposons au contraire l'existence d'entiers p et q tels que $\sqrt{2} = p/q$ et qui peuvent être supposés sans diviseur commun. Alors

> $p^2 = 2q^2$...", et on aboutirait à une contradiction. Les variables p et q sont liées existentiellement dans la première phrase, et on les prend libres dès la suivante; si, dans la première phrase, on remplace les variables p and q par d'autres variables, les variables de la seconde auront perdu leur support...
> Ce qu'il faut dans tous ces cas, c'est annoncer explicitement que les variables liées existentiellement seront désormais employées en tant que noms propres d'objets qui remplissent la conditon imposée.

or 20 ans ont passé et cela n'a vraiment pas changé du tout!

Dans son désir de lutter contre toutes les scléroses et d'aider au progrès partout dans le monde, je pense qu'on peut dire que c'est lui qui est à l'origine de la reconnaissance et du développement institutionnel de la didactique des mathématiques en France puisque c'est lui qui, un jour de 1975, m'a déclaré qu'à son avis les travaux que j'avais faits sur l'enseignement des mathématiques constituaient une thèse, ce qui, à cette époque, était inimaginable en France (aux Pays-Bas, il avait déjà depuis plusieurs années fait soutenir à Van Hiele une thèse sur l'enseignement), mais l'autorité scientifique de H. Freudenthal a été déterminante pour convaincre une commission de thèses de mathématiques en France de commencer à prendre en considération des travaux de ce type. Je ne ferai pas ici la liste des didacticiens qui dans de nombreux pays ont éprouvé des difficultés analogues et n'ont eu le courage de persévérer que grâce à l'estime de H. Freudenthal. Nous lui devons tous une immense gratitude: sa personnalité était si extraordinaire que non seulement il nous inspirait respect et admiration pour lui-même mais il nous conduisait aussi à une plus grande assurance en nous-mêmes en voyant que nous avions gagné sa confiance.

Certes, parmi ceux qui ont travaillé sur l'enseignement des mathématiques, certains se sont opposé à lui, même violemment (G. Howson raconte que Freudenthal répondit un jour à Dieudonné: '*Don't shout at me, because I can shout as loud as you can and in more languages*'), mais tous ont dû reconnaître l'importance de sa pensée. Cette pensée ne périra pas puisque (cf. '*Weeding and Sowing*'!) il a arraché beaucoup de mauvaises herbes et il a beaucoup semé.

NOTES

1 Cet article reprend et développe un article que j'ai publié sous le titre '*Hans Freudenthal (1905–1990)*' dans le bulletin de l'Association des Professeurs de Mathématiques de l'Enseignement Public – no. 378, avril–mai 1991.

2 '*L'origine de la topologie moderne d'après des papiers inédits de L. E. J. Brouwer*', avril 1975 – publication DD3 de l'UER de Didactique des Disciplines de Paris 7.

3 '*L'échec des coureurs*' – Congrès APMEP – Lille septembre 1983 – Bulletin de l'APMEP no. 342 – février 1984.

4 '*Erreurs du professeur – Analyse didactique de soi-même*' in Actes de la rencontre CIEAEM 39 – Editions de l'Université de Sherbrooke – 1987, pp. 37–41.

5 Débat sur la Réforme de Mathématiques Modernes entre Madame Krygowska et Monsieur Freudenthal – in Chantiers de Pédagogie mathématique – Régionale parisienne de l'APMEP – Cahier 33 – juin 1975.

RÉFÉRENCE

G. Howson: 1985, 'Hans Freudenthal and the foundation of a discipline of mathematics education', *Nieuwe Wiskrant* **5**(1), 68–72.

Institut des Sciences et Pratiques d'Education et de Formation
Université Lumière Lyon 2
16 quai Claude Bernard
69007 Lyon France

It took me four hours to write down these thoughts (filled in later) but during those four hours I allowed myself to be distracted by a hundred other things. I got up to go to the window when someone passed by on a bicycle. Without getting up I watched the neighbours across the street who themselves were looking out the window. I paced the floor, much to the displeasure of those below. I practised what one might call the art of singing, if that term did not already have a definition. I took the atlas from the shelf three times, not because I wanted to look up something geographical, but because an atlas is such an attractive book. But besides the atlas I also perused several other books. Of course I filled my pipe several times, lit it, cleaned it. And finally I put on and took off my glasses several hundred times – the most enjoyable and most versatile relaxation that has befallen me in my climbing years thanks to the corresponding presbyopia. This is how I do scientific work. Mathematics is a demanding profession, one I find it impossible to concentrate on. Two minutes of mathematics at a time is probably the maximum I can perform. Under, to the side, on top of the sheet of paper with formulas there are books, magazines, photos, catalogues, and notebooks which surround and enhance the mathematical activity. There are other characters, but this happens to be my way. I have not divided my time between work and pleasure, but have spread out both homogeneously over the day. Like in a solution, of which I could really not say how much there is of one ingredient and how many of another. This is my way of life, with no discontinuities, without separation lines between work and relaxation, nor between the various kinds of work and relaxation.

From: H. F.: Vrijetijdsbesteding, Wetenschap & Samenleving **14** (1960) 77–80.

FRED GOFFREE

HF: WORKING ON MATHEMATICS EDUCATION

ABSTRACT. First an impression will be given of the way Freudenthal worked on mathematics education during the period after WW II, until the early seventies. The sequence of events of that time form the leading principle of that period when important didactic concepts were being developed. In the following section the chronological order is set aside in an attempt to show how Freudenthal approached research, both in education and into mathematics education. In the next two sections two core themes are dealt with in detail: 'learning mathematics' and 'developing mathematics education'. Earlier didactic concepts are completed and further developed into key notions of a 'science of mathematics instruction'.

INTRODUCTION

To develop mathematics education, there are three possible lines of approach, namely by being interested in education, by thinking about mathematics education and by contributing to educational development. This would normally be a full time job for three people if it is to be done well, but Professor Freudenthal seems to have done it 'on the side'. One would expect to be able to put into words what this man has achieved. This, however, proved a very difficult task, and the extent of his work can never be fully covered. However, I knew this when I started for, as a colleague working in his close proximity, I saw the list of his writings grow during twenty years. Therefore I decided to make a selection from articles, papers, newspaper cuttings, notes and books as the basis for my article[1], my first object being to do justice to Freudenthal. But an article has certain limits and the author has his limitations, therefore this is no more than a sincere effort. Please judge for yourself.

1. MATHEMATICAL DIDACTICAL NOTIONS IN DEVELOPMENT

HF

Early in July 1981, HF had finished an important productive period. The book of which staff members of IOWO had seen the first Dutch chapters years before, had now been translated into German as well. The English version appeared in 1983 at the publishing house Reidel in Dordrecht under the title *Didactical Phenomenology of Mathematical Structures*. It was a tremendous contribution to the development of mathematics education – a thick book in which the constructive element dominates and which shows us HF as a designer as well. This will be dealt with later.

Educational Studies in Mathematics **25:** 21–49, 1993.

What to do next? Slightly bewildered, he must have asked himself this question, for besides his own initiatives there had always been requests for contributions to conferences, scientific journals and other media, which prevented him from choosing what he wanted to do next for an extended period of time. At least this is how he saw it when he started his preface to a new book, a completely different one this time, covering the story of his life. It was not to be an autobiography for he failed to see (or would not see) a clear structure in that but rather bits and pieces, snapshots from his life, gathered while cleaning out his bookcase. The result appeared in 1987: *Schrijf dat op, Hans* (Make a note of that Hans) published by Meulenhof Informatief. Those who had known HF during the preceding years, like I had during numerous train journeys to Enschede, where he helped the mathematics department of SLO to get started, would from time to time hear a few fragments. He wrote the way he talked. No, actually it was different, what he wrote sounds like a story, very spontaneous and told in a relaxed way. This was not only the case with his anecdotes and stories, but also with his crystallized thoughts on education[2].

On page 66 there is a chapter headed 'H.F.'. This contained a modest curriculum vitae, as leitmotif for the readers. The initials H.F. had in earlier years acquired a meaning of their own for his staff. Prepublications of articles, little problems or cuttings, signed HF, would regularly circulate at IOWO. HF'75 and HF'80 stood for exuberant parties celebrating his birthdays in 1980 and 1985. This seems ample reason to continue the tradition in this article, so HF it is.

Only on page 335 did HF reminisce about his experiences with education. This chapter was titled: '*Working on education: on our way*'. Other highly qualified scientists began to pay attention to educatonal matters as they got older, but HF was attraced from an early age. He explains this very simply: 'All my life I have been a poor teacher, and in order to make the best of it I started thinking about educaton at an early age' he said. Thus learning and teaching became subjects for reflection and study and the results of all that reflection which was increasingly inspired by his own observations in practice and greatly stimulated by interaction with practicians and theoreticians from all over the world, made HF internationally famous. How his thinking developed in the course of half a century HF failed to understand. This is not surprising for that reconstruction would have to be very thorough, like his work, reflective and down to the smallest detail of the personal learning process. It would have been interesting to find out where exactly new ideas had come from, what helped generate them, when, how and in what context did they develop? However HF did not want to describe these memories at the time. He did so a few years later in his last extensive educational book that appeared in the series 'Mathematics Education Library', under the title *Revisiting Mathematics Education. China Lectures*. Kluwer Academic Publishers, 1991. The first copy was presented to the Freudenthal family by H. Bauersfeld on the occasion of the opening of the Freudenthal Institute in Utrecht.

Prehistory

Before we come to this last book, we still have a lot of ground to cover. Therefore we shall go back in time to the period that HF usually called his prehistory[5]. We return to the twenties in Holland, to 1924 when a supplement branched off from Het Nieuw Tijdschrift voor Wiskunde (the New Journal for Mathematics), which later became *Euclides, monthly journal for the didactics of mathematics.* In this magazine would write experts in the field of mathematics education such as E. J. Dijksterhuis, who reacted in 1924 in the first supplement to a pamphlet by Tatjana Ehrenfest-Afanassjewa about the use of geometry education for non-mathematicians. The subject under discussion has become dated, for geometry education, as it was called at the time, no longer exists. However, HF found the underlying idea positive, in contrast to his reaction to a curriculum plan, made by the Beth Committee. HF believed he saw the signs of an ambiguous view against which he was to fight continuously, namely that the purpose of mathematics education was mainly to 'contribute to mental educaton and development'. So the educational value of mathematics education was its main aim. It is easy to guess what was behind it. The material for mathematics is so well-organized and its structure so clear that this alone should instill in pupils a logical way of thinking. Later HF was to connect this piece of prehistory with something that E. W. Beth, the son of the aforesaid H. J. E. Beth (of the committee) said, more than thirty years later: the aim of mathematics educaton is mainly becoming acquainted with the 'deductive method'.[6]

A College of Mathematics Education

What happened before 1930, when HF came to The Netherlands, around 'Euclides', 'Ehrenfest' and the 'Beth committee' had little or no effect on practical mathematics education. But on April 20, 1936 the Mathematics Study Group of the WVO was founded.[7]

During the weekends there were discussions among the members at their homes about the essence of mathematics education in secondary schools. From 1948 onwards the larger themes would be treated during Autumn conferences and from 1948 till 1952 hard work was done on a mathematics programme for secondary schools – a design of the Mathematics Study group of the WVO.

But we can't skip the period of WW II. For during this time HF could devote himself to elementary mathematics (in other words primary school arithmetic). The lessons in arithmetic that he gave his two sons at the beginning of the war inspired him to study anything that had been published until then in this field. Booklets about arithmetic, rationals of schoolbook series, didactic theories, research in the field of arithmetic education, work by important educators and pedagogues. Whatever he came across was filtered by his thorough knowledge of the history of mathematics, observation of his childrens learning processes and his conviction that (new) mathematical knowledge could never be acquired by deduction. At the time he was already capable of distinguishing between good and bad, interesting and worthless, original and copied and well-thought out and misunderstood, although he was not connected with training at a school. He wrote down his findings around 1942. If you go to the Public Record Office of Noord-Holland at Haarlem, you can read them.[8]

It is the first chapter of a 'didactics of arithmetic' typed out on 103 folio sheets. HF acknowledges however that he never got round to the actual arithmetic didactics. When one reads the manuscript it is rather as if he tried to free the (experimental) garden of arithmetic education from all weeds. He discusses in particular all that is bad, worthless, copied and misunderstood. The metaphorical sowing that should follow all this weeding had to wait until the well-known publication of 1978: Weeding and Sowing. But whoever is looking for the roots of important ideas that HF launched and elaborated in the course of half a century, will be absorbed by reading his fragment about the 'didactics of arithmetic' dating from 1942. It was not difficult to find a number of interesting indications while merely leafing through the 103 typed folio sheets. First a few about HF's method:

- little regard for advanced statistical calculations with simple data and reports of psychological research in the form of tables;
- mistrust of quoted research and always looking for the original sources;
- strongly recommend the use of a journal/log by teachers;
- a warning to didacticians who think they can find ready made tools in other fields of science for the method of arithmetic education;
- the conviction that the applicability of a pedagogical method can only be shown in practice.

Next there are a number of points that were to be further explored or would be embodied in later writings:

- the difference between 'number concept' and 'the concept of number';
- the idea at the time to teach in little groups of 3 to 6 pupils;
- the (arithmetic)didactic preparation of teachers;
- the subject matter, apart from the activity of learning, never interested HF sufficiently to devote special attention to it;
- organizing subject matter.

'Every positive action begins with criticism. That is the meaning of this meaningless chapter', apologized the author on page 91 before starting on the next chapter. That was to deal with the aims of arithmetic education. The beginning was promising, but it was never finished. This first product of his pen on education was abruptly broken off on page 103. The sudden end probably had nothing to do with the war nor with the resulting circumstances, and even his stay at a camp in Havelte[10] (1944) could not stop HF from thinking about mathematics and education (and various other fields) (Figure 1).

WVO held its first conference after the war at the castle Rhederoord in De Steeg, from 11th to 14th August 1945. Afterwards HF wrote 'Struggle for Life' praising the solidarity of the participants and their cooperative attitude. It must indeed have been a special experience compared with the attitude of 'every man for himself' that was common during the war.

> It is less than 6 months ago that people who had certificates, diplomas and degrees were all like sparrows fighting for crumbs in the street and in the daily

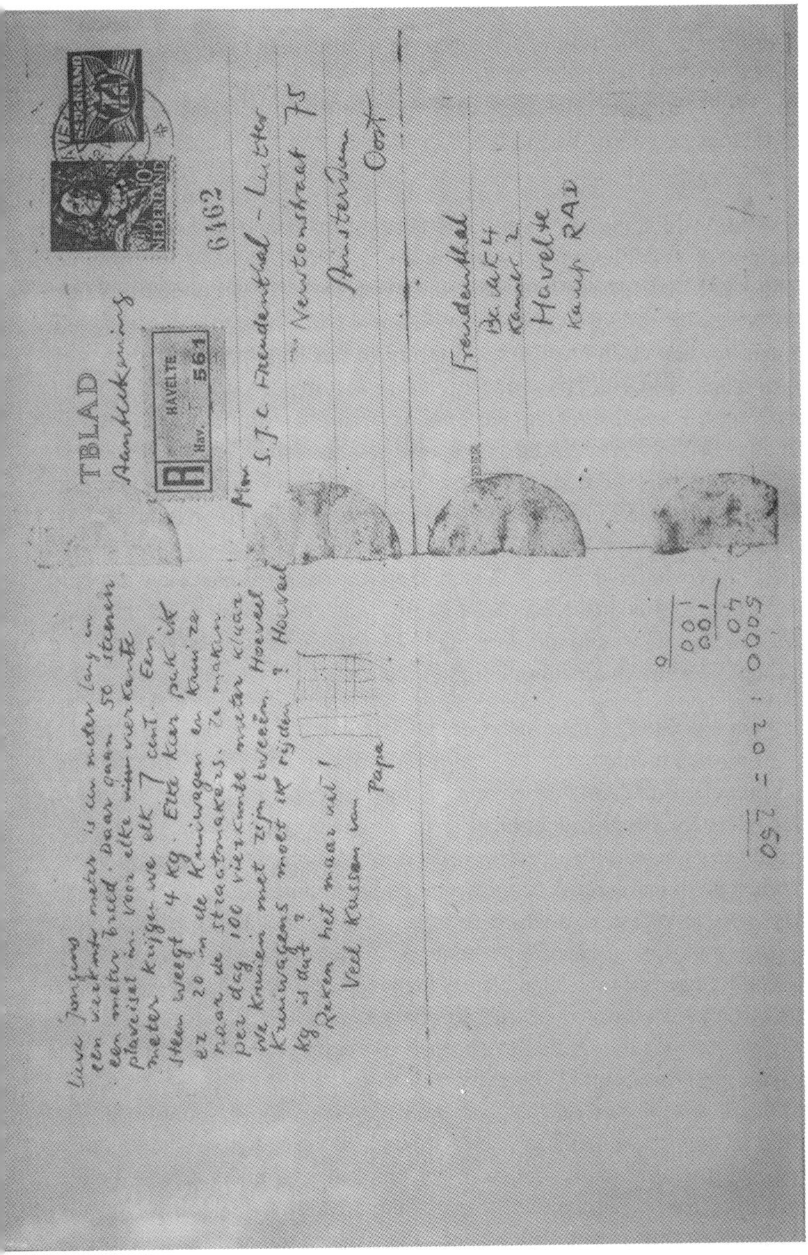

Fig. 1. The postcard from Havelte.

> battle on the steps of the tram platform little has changed from the primitives we were then (...)

At the conference itself he spoke about a subject that had occupied him for a long time in view of the fragmentary '*Didactics of arithmetic*' of 1942, and with which he had not yet come to terms. The title of his lecture was: '*Education in thinking*'[11] and the following passage reveals some of the struggle with this subject matter:

> (...) Education in thinking – isn't this a beautiful sounding and attractive combination of syllables: what a tempting task! I would not say this so pertinently had I myself not been seduced by them, had I not done this same work for years on end – or put more modesty – had endeavoured to do, full of enthusiasm, just as enthusiastic as I return to this task today after years of unemployment. However I have never felt safe with my task and the realization of its necessity. For whoever teaches reading, writing, arithmetic or French or mathematics or physical training or whatever other subject, knows what he is teaching even though he does not know how to do this exactly. Thinking is not a skill. What is thinking actually – I have often thought long and deeply about this but every time I tried to find out something more about this curious 'thinking' that I wanted and had to teach, then I found I had no idea whether it was I who went around in cirlces or the world I wanted to observe, the thinking that was dancing all around me. Gradually I have learned to see this problem of thinking in a more practical rather than theoretical way. (...)

The Mathematics Study Group also went back to work in June of the following year. They discussed mathematics education and tried to develop curricula. The way they approached mathematics education was the opposite of what the Beth Committee and its followers had advocated during the previous years.

Mrs. Ehrenfests '*Uebungensammlung zu einer geometrischen Propädeuse*' from 1931 must have had a positive influence on the forming of opinion in those circles. HF was to praise it often, also much later.[12] Yet he had his doubts about the position of this 'geometric foundation course' in the whole of geometry education. For Tatjana Ehrenfest had merely made the entry into the solid structure of geometry easier. And it was this structure she really cared about, for the subject of geometry was an ideal exercise to sharpen one's intelligence, because pure reasoning was least interrupted there by day-to-day diversions. In 1951 HF publicly took issue with her on this last point.[13] Geometric orientation in the world around us can be valuable in itself. It teaches one how to organize one's surroundings with simple mathematical means and thus to understand it better. So geometry becomes a 'rich context' for mathematical activities. But HF only began to think along those terms much later. The work in the Mathematics Study Group inspired and moulded HF to a great extent. He called it his 'College of mathematics education' referring to the monthly meetings where he got the response to his creative ideas and productive mind so badly needed in order to go on developing.

Fig. 2. 1949 Maarten Maartenshuis in Doorn. HF surrounded by mathematics teachers. The photo was taken by Pierre van Hiele.

Learning Processes

During the period HF was coaching his PhD students Dieke and Pierre van Hiele his thinking about learning processes rose to new heights. Dieke described in detail how her pupils in the lower part of a secondary school were taught geometry and Pierre's earlier suspicions were confirmed by her records. Thus his 'level theory' came about.[14] He found that learning processes are discontinuous, for pupils clearly showed jumps in their efforts to acquire insight. This caused Van Hiele to deduce that there are levels of thinking. On a lower level things are done intuitively, informally and acting with objects, but this is regarded differently at a higher level.[15] These acts of a minute ago are the objects of mathematical analysis. But HF has filled in the levels differently and he attributed the rise in levels (later) to reflection. We shall come back to this in the paragraph about 'learning mathematics'. The research of the practician Dieke van Hiele-Geldof and the theoretician Pierre van Hiele had a tremendous influence on HF which he freely acknowledged.

Modern Mathematics Education

HF had initially directed his attention mainly to secondary education. This was partly due to the Study Group but also to other circumstances. However he had a wide outlook, which was also stimulated by his column in 'De Groene

Amsterdammer'[16]. Through essays he brought many educational problems, both big and small, to the attention of a large audience. *Big* problems such as '*Education to become complete human beings*' of 21 April 1956, an impressive picture of education in Southern Asia.

> (...) This was the only place where I could really experience the feeling that the more serious problems of education and pedagogy are the same all over the world. There was no struggle here with an old-fashioned system but here was the humbling peace of one who is seeking and knows he is on the right track. My best memory of Bombay is the morning service at this school which was held in a bare large room that could only hold half the children – five hundred out of one thousand. One soon gets used to large crowds in Asia. The swarming at the bazaar, the multitudes celebrating Sunday on the walls of the quays and the steps of the jetties, one gets used to all those races (...).

And *small* problems as in '*Practical Arithmetic*' of 14 October 1950.

> At table Miriam, who is rather proud of what she learns at school, asked Tom: "Twenty half pennies. How many pennies do they make?" Tom not willing to be outwitted by his younger sister answered: "Half pennies no longer exist." And Miriam not to be outdone said: "But these sums do" (...).

In the last but one of the articles in this series, on 28 July 1962, another large, comprehensive and extremely topical problem came into the limelight. HF wondered what the new education act (in Holland nicknamed the Mammoth Act) was going to look like (in the classrooms). Can laws and Royal Decrees really change education? No they can't. Do textbooks and exam programmes influence education in class? Yes, rather. But the teachers and to a certain extent the pupils finally determine how things are actually done. This opinion was very up to date, for a year before the Modernization Committee of the Mathematics Curriculum (CMLW)[17] had begun the introduction of Modern Mathematics in Holland by first reorienting the secondary school mathematics teachers. The introduction of this new form of mathematics could only succeed according to HF if the teachers at least had mastered the subject matter sufficiently. HF openly declared himself in favour of a new curriculum based on Modern Mathematics, which he had been the first to introduce to the Dutch lecture rooms. But what did not become clear in the newspaper article and became known all over the world only a short time afterwards was that HF did not mean New Math but modernized education in mathematics: which is an entirely different matter.

However more was to come. When the CMLW was installed HF was absent because he was in the United States for a sabbatical. On his return in the Summer of 1961 he found the minutes of the first meeting. He was extremely disappointed because they had merely concentrated on programmes for the Gymnasium (Grammar School) and HBS-B (former Dutch High School for 12 to 18 year-olds) and not a word was said about improvement of mathematics education. He showed his disappointment in a letter to the chairman of the Committee, Prof. Dr. A. F. Monna dated August 25.

> (...) I have said several times that I do not see the need for the modernization of the curriculum as propagated by many. Not because I dislike modern mathematics but because in the various proposals the introduction of modern subject matter is seen as the prime objective. The only urgent matter I can see is the improvement of mathematics education. (...)

A result of this intervention was that during the following years the interest of the CMLW and its subcommittees would also be extended to MAVO (school for lower general secondary education) and LBO (lower vocational education). Teachers from those sectors were given in-service courses both for the new subject matter and for a different didactical method.

'How to Teach Mathematics...'

Internationally HF also put forward his ideas with great zest. From 1954 he was the chairman of the Dutch Educational Committee for Mathematics and soon after he became a prominent member (president in 1967) of ICMI, the International Commission on Mathematics Instruction. Thus he was one of the first to see the beginning of the New Math movement. Oddly enough he underestimated its impact at the time, as many did with him. Maybe the proposals for the introduction of new subject matter (sets, logic, relations...) were too extreme, too much in conflict with what he had recently proposed himself for him to take them seriously. But the reason he stayed away from the conference, held in Royaumont in 1959 under the auspices of the OECD, which prompted the New Math movement in Europe, was a different one. Too many mathematicians, who had nothing whatsoever to do with mathematics education, had been invited, according to HF. He did not want to swim against the tide again. Later he came to regret his decision which he mentioned in various ways and in various places.[18]

From 21–25 August 1967 a symposium was held in Utrecht with the title: 'How to teach mathematics so as to be useful?' It is not hard to guess who wanted to discuss this theme at an international level. HF grasped the opportunity to explain his views, which had earlier been presented to a Dutch audience in his column, at an international forum: Modern Mathematics is, proportionally speaking, a slight amount of knowledge with which you can do proportionally much more. Or, in other words, the more abstract mathematics is, the more applicable it can be. Abstract means: loose from its context. The power of mathematics has always been that the contexts in which the mathematics had been created could be eliminated. But this HF said was exactly the problem for mathematics education.

> (...) Take such problems as:
> If I have got ten marbles and I give three away, how many are left?
> If I have got ten marbles and John has three less, how many does he have?
> If there are ten students in the room, and three are girls, how many are boys?
> If I am ten years old now, how old was I three years ago?
> If B is between A and C, B is at a distance of 7 miles from A, and C is at a distance of 10 miles from A, how far is B from C?[19]
> It is not easy to learn that in all these and a hundred other situations the

> same arithmetical operation applies. It takes some time, but finally everybody succeeds in understanding it. Why? I daresay, because arithmetic starts in concrete context and patiently returns to concrete contexts as often as needed. (...) Arithmetic and geometry have sprung from mathematizing part of reality. But soon, at least from Greek antiquity onwards, mathematics itself has become the object of mathematizing. (...) Systematization is a great virtue of mathematics, and if possible, the student has to learn this virtue, too. But then I mean the activity of systematizing, not its result. Its result is a system, a beautiful closed system, closed, with no entrance and no exit. (...) What humans have to learn is not mathematics as a closed system, but rather as an activity, the process of mathematizing reality and if possible, even that of mathematizing mathematics. (...).[20]

Although it was yet invisible to many, including the author of this article, a number of important 'didactic notions' were already put forward in this lecture: *mathematics as an activity, mathematizing, context.* Insiders even recognize what was later elaborated by Treffers and was included by HF in his repertoire with diffidence: mathematizing horizontally ('mathematizing parts of reality') and mathematizing vertically ('mathematizing mathematics itself').

The opening address and contributions from other leading experts in the field of mathematics education[21] were published in the first issue of a recently founded international scientific journal for mathematics education: *Educational Studies in Mathematics.* Founding editor H. Freudenthal.

Conceptual Tools

Slowly the seventies are coming closer and with that the institute that enabled HF to observe how his ideas, concretized in material for pupils and teachers, were brought to life in classrooms: the IOWO, Institute for the Development of Mathematics Education. So it was not an institute for curriculum development. HF had chosen the name with care because he was concerned with the *development* of education: from nursery school up to and including grammar school, for pupils and their teachers, carried out at school and with all those concerned. Publications after 1970 made this very clear: what was designed by the staff of the IOWO, discussed, tested, changed, described, presented etc. was an endless source of inspiration to him. (From the start HF felt one with them: neither greater nor less). An endless stream of papers, lectures and articles was written by him, only interrupted by four major works on mathematics education.[22]

It was not easy to insert a leading principle in the above that would enable the reader to see some development (of ideas, of mathematics education, of the formation of theories, of HF himself). But this becomes even more difficult in the period after 1971. HF shared this opinion. He did not like the idea of applying some structure or other, with the usual abridgements, to his own process of development. In the preface to a simple collection of articles 'Apples and Pears/mathematics and psychology'[23], he wrote:

> He who is successful as an author, will have to choose from a diversity in order to combine essays into some sort of unity. The red thread running through the parts of his work is actually too long and it has too many branches to put it through the small number of beads.

Therefore the choice of red threads in the remainder of this article has been made by the author himself. It is clear from the start that it will never cover the scope of HF's contribution to education. However I hope that it will give a good impression of the important 'didactical notions' the roots of which have been indicated above. They will become key notions of a theory of mathematics education. To this I should immediately add that HF would almost certainly have replaced 'key notions' in the context of didactics by 'mental objects'.[24]

However the author of this article considers them as the building blocks for a theory 'of and for' mathematics education. Let's have a look at what building materials were available at the time. It will simply have to be a mere listing: *structure of the subject material, the problem of thinking, mathematics as an activity, rich context, mathematizing, discontinuity of learning processes, raising levels and reflection*. The basic material was there, a rich basis of orientation had been created for future developments. The way it was to be used was to generate besides deepened insight and larger connections also new building blocks. HF produced a number of these out of his hat in his last book as 'conceptual tools': *phenomenological approach, mathematization in all its forms, context, inventing afresh under guidance, levels of learning processes, paradigms of learning, reflection, insight, anticipating learning and learning by reflection, intertwining learning strands, mathematical attitude*. 'And many besides', he added.[25]

2. INTERLUDE
RESEARCH IN EDUCATION: A STANDING JOKE AND A FEW RAYS OF HOPE

The Captain

On a boat there are 26 sheep and 10 goats. How old is the captain? This question was put to 97 children between 7 and 9 years old by researchers of the IREM at Grenoble around 1980. Of the 97 children 76 managed to calculate the captain's age on the basis of the data. They probably added up the numbers given. One can be amazed at this and the French researchers were. Therefore they decided to continue their research with a larger number of similar problems which were now also put to older pupils. At the end they also asked what the pupils thought of these questions.

Oddly enough there were children who went to any lengths to find the answer on the basis of the data given, but who said afterwards that they found the questions stupid. This fact was the reason for the researchers to attribute this behaviour to the school culture where sums can always have one correct answer only. But HF looked somewhere else for the cause viz. in the 'magical context' which is often evoked in children and which had not been recognized as such by the researchers of arithmetic education. The above is described extensively in an article in *Educational Studies in Mathematics*[26]. The reader will also notice that 'the captain' is preceded by a passage about the application of the Kuder–Richardson KR20 test in educational research. Now HF was connecting the two in order to denounce educational research in his particular way. This time he was mainly concerned with the largely misunderstood application of statistical methods. Whoever believes that he can understand how children learn by means of e.g. the so-called regression method, is no better than the children who on the basis of the number of sheep and goats on board can calculate the age of the captain. This was the moral of the story. For those researchers however there is something called the magical context...

This section is called interlude. It is intended to simplify the reader's transition from the previous chronological order to the more thematic writing we shall now adopt. The topic of research is ideal to mark this transition. HF was very active to this point. In particular, during the early seventies he tried to expose the failings of educational research (and educational theory) in many respects. Particular researchers and reports became the object of his criticism. He was very good at this. In Weeding and Sowing of 1978 he summarizes his criticisms and organizes them systematically in the chapter 'On education'. What was HF aiming at? At atomizing subject matter as an undesired result of behaviourism. At the new wave of goal setting; the formulation, classification, operationalization and such as an educational game that was useless for education. At the childish belief that all knowledge is organized according to concepts, and that knowledge is acquired by forming concepts. At the idea that we surely can learn these things from inquiries and questionnaires. At diagnostic tests designed for Mastery Learning by 'doubly blindfolded' test constructors.[27] In particular at the multiple choice form which required reading ability and logical thinking rather than knowledge which is supposed to be measured in this way. At the way in which curricula were developed by generalists and based on empty models that had never yielded successful results in class. At the over-simplified way models were used, which sometimes had the nature of a pre-view but are often more an after-view. His criticisms were far from mild, occasionally even sarcastic in tone and often larded with witty puns and funny anecdotes. When HF denounced the abuse of mathematical models for instance, he showed with great pleasure the following monstrosity that he had made himself:

A painting is an ordered threesome [R, C, I] consisting of:
a rectangle R in the Euclidean plane;
a set C the elements of which are colours;
and
a relation I between R and C, where I(x, y) means: in point x the colour is y.

Looking for Paradigms

Meanwhile the challenged research simply continued. Just like HF did with his IOWO. There he had seen a ray of hope during all the criticism in the form of so-called development research[28]: no momentary impressions of learning or cross sections through learning processes, but following pupils during a longer period of time. Not trying to collect data about lots of pupils in order to come up with general laws of learning via statistical techniques, but describing pieces of information from the learning process of, e.g., one pupil. In particular paying attention to the moments when new insight is created in the pupil, with the hope that the researcher will observe something worthwhile about the learning process and level raising.

> (...) I will try to make clear what I mean by observation of learning processes by means of an example: I always spoke with Bastiaan (born on 27–4–1970) in adult language, sometimes purposely introducing words that I supposed were new to him. Whenever he did not understand he would say: 'What did you say Grandfather?' and then I would repeat what I had said previously but in different words. On 15 February 1975 he did not say 'What did you say Grandpa?', but he said 'What does 'elsewhere' mean?' This proved to be a watershed for never again did he react to an unknown word or construction with the words: 'What did you say Grandpa?' but he would always ask about its meaning. (...) I was not helping him, I could not say whether this was on purpose or not. But at any rate I enabled him to reach a discontinuity in his learning process, a learning jump. I regret however that I did not draw his attention to this fact at the time: he had replaced the global question about the meaning of a linguistic utterance by a much more direct question about a word (...).

Later he added in the same article:

> One learns more from one paradigmatic case than from a hundred that do not apply. This is how you should interpret my observation of 15 February 1975. The situation described by me is a unique one, in particular when I tell the previous history as well. It is a windfall when one can observe a caesura as clear as this one, but one should make the most of the opportunity too...

The word paradigm has been used. HF also used it in relation to learning. There paradigms are examples that provide the pupil with new insight. One example is then sufficient to be able to apply the acquired insight elsewhere as well. Paradigm is a concept with a form of learning that is indicated by the word 'exemplary'. He also used it in a didactic sense. A researcher tries to find paradigms, as for instance in the case of Bastiaan described above, that enable new didactic insight. It is no longer necessary to do hundreds of similar observations to legitimize the insight. Paradigms are the rays of hope in the lives of researchers such as HF. Development research as carried out by the IOWO and defined there due to the fact that it was to be abolished at the end of the seventies, combined all the rays of hope into one shining beacon. OW&OC that replaced the IOWO from 1981 onwards could orient itself by this beacon.[30]

B. S. Bloom

We will return for a moment to HF's criticisms on educational research. We should use this opportunity to turn to the concept examplary for help. For this will enable us to take a closer look at his criticisms without having to mention all the papers HF published in this field. In '*Weeding and Sowing*' HF extends a helping hand. Paragraph 2 of the chapter '*On Education*' is entitled: '*In Full Bloom*', and as I write this I can still remember his merry eyes when he presented us with this pun. The criticism that HF poured over Bloom's head, may undoubtedly be called exemplary for his criticism 'on education'.

His first contact with work of B. S. Bloom was purely coincidental. In his capacity as editor of various magazines, HF had often been annoyed by the way certain sources were referred to in articles on education. A name and a date were given, complemented in the bibliography with a title and that was all.

HF thought it was not sufficient and it induced him to ask himself a few questions about the relevance of such references. Had the sources indicated anything to do with what was put forward by the article? Had the references been quoted for support? Were they complete? Consistent? What had been done about sources that refuted the presented research? Etcetera. He started a project on this field, but it proved far too ambitious and could therefore not be carried out. Then HF came across '*Human characteristics and school learning*' by B. S. Bloom (1974). Here praises were sounded about the system of Mastery Learning that determined the educational fashion in the mid-seventies.[31] It was not merely praise, there was evidence too that would show that the system actually 'worked'. HF would not mind seeing that for himself, especially because he had found a few remarkable inaccuracies in Bloom's book already. And thus he went in search of the source of this evidence which turned out to be unpublished dissertations from the University of Chicago, where Bloom was working. HF sifted through 5 dissertations and found that in the first place they had not been quoted correctly by Bloom and secondly that those researches left a lot unanswered. All had used mathematical subject matter (e.g. matrix algebra and probability), but the researched education had been too short and of poor quality. Thus 'mastery' was contrasted with 'poor education', the results of the diagnostic tests were not analysed and moreover the smallness of the random checks made it impossible to draw any relevant conclusions. So it was one big mess, a witches brew which HF stirred around with relish for 30 pages.[32] It is not certain whether he was really angry about such abuses, but he would not beat about the bush, as can be seen in the following passage:

> After this wave of criticism the reader is entitled to ask whether I did not joke when I introduced this work as an impressive study. No I did not joke. It is impressive, like a huge organ, with all registers pulled out, or a famous orchestra hired to play 'Baa, baa, black sheep, have you any wool?' It is impressive as a show of instruction raped by instructional research.

Mastery Learning occupied him for some time. In The Netherlands too he would meet projects that pretended to have proven the superiority of this new system. A fine example is a research study at Enschede, subsidized by the Dutch Government and carried out by a local Pedagogical Centre. The aim was to compare a new approach to arithmetic teaching that had been set up according to the ideas of Mastery Learning with arithmetic education not using this system. According to HF the research report lacked a great number of essential data. HF asked the project leader 37 questions and received poor answers to most of these and to some he got no answer at all. An exchange of letters followed (somewhat one-sided) but this yielded nothing. After 6 months HF published in *Pedagogische Studiën*.[33] In our country he did not remain silent either, one sentence illustrating this:

> The report on the effective aspect fills the greater part of the final report: correlations, variant analyses and multiple regressions – a nightmare of figures.

Bloom had not become well-known due to Mastery Learning but because of his *Taxonomy of Educational Objectives. The classification of educational goals. Handbook I: Cognitive Domain*' which drew a lot of attention in the seventies and was frequently quoted in just about every educational book. His taxonomy was a hierarchical scheme consisting of 6 layers: knowledge, comprehension, application, analysis, synthesis and evaluation. In these layers the learning goals of a certain subject matter could be arranged. An earlier mentioned category in this structure was considered conditional for one that was mentioned later. Therefore 'knowledge' came first, and after that 'comprehension' so first you have to know and then you have to understand. HF realized three things at once. In the first place the six categories of taxonomy could not be applied to the learning goals of subjects such as mathematics and physics. 'Observation', 'experimentation' and 'designing experiments' did not occur for instance. Secondly Bloom and his coworkers showed in their examples that they had not interest whatsoever in learning processes and levels. And thirdly this emphasis on goals was dangerous. HF indicated this danger with a surprising change of perspective simply like this:

> It is most astonishing, however, that in the classification something is lacking and in no way to be placed, which could be called the ability to pass tests, that is, reacting adequately to them, a complex ability in which partial abilities can be distinguished such as:
> insight into test structures;
> ability to disentangle test structures;
> ability to weigh evidence;
> insight into the psychology of test procedures.

That was the way matters were going, learning goals could not describe everything worth teaching and learning, but education would be restricted to what was put down in the learning objectives. In other words: first try to test what has been taught, later they try to teach what will be tested. This is disastrous in mathematics education. HF had seen examples of that in the past. Was that the reason why he mopped the floor with the efforts to put objectives of mathematical subject matter into the taxonomy?[34] This seems more than likely. Did it achieve anything? We shall never know, for the interest in 'Bloom's taxonomy' had to a great extent disappeared at the end of the seventies. Other phenomena claimed our attention. So much for Mr. Bloom on whom HF spent a great deal of energy.

Piaget

HF devoted even more energy and words to a researcher who not only gave him food for thought but also inspired him to do tests: Piaget. HF greatly valued Piaget's observations of his children, but he fought his later more theoretically oriented work by fire and sword. In many articles and also in his books HF explained that Piaget was often wrong with his experiments and his theoretical analyses. However in later comments the firceness seems to have abated a little. HF was very much aware that Piaget took the point of view of an epistemologist in his educational work. This implies that he was occupied by knowledge and its structure in the human mind. Piaget was not a didactician and what he put forward about cognitive structures could not be seen as a contribution to didactics. Unfortunately however, this often happened.

So what was HF's main objection against Piaget? Well firstly the publications about the development of number in children. HF undoubtedly read the book by Piaget and his assistant Alina Szeminska 'La génèse du nombre chez l'enfant' published in 1941, from cover to cover. From this it became clear to him that amongst other things, Piaget had not considered the influence of language during his clinical interviews with individual children. Therefore interesting observations were misinterpreted. But what really concerned HF was that this book did not in fact deal with the 'development' of number in children at all. The basic assumption of epistemology is a knowledge structure as it has been built up in human beings. In this case knowledge of (natural) number. Piaget tried to discover how children of various ages understood this 'number'. He put children in artificial situations and asked them rather unusual questions. As was said earlier he did this without paying attention to the language used. Moreover the knowledge structure of natural number that had been chosen was one of the most advanced of that time. The cardinal aspect of number dominated, and counting was out of the question. Later during the period of New Math, others could claim Piaget's authority in order to start teaching numbers according to Cantor's idea of equipotent sets at nursery school. One drew sets and lines in varius colours were drawn from one element to the other in order to show the $1-1$ correspondence. But counting was out. This disturbed HF greatly and also the fact that Piaget thought that the cognitive development in children took place from poor to rich structures whereas HF thought it was the other way round. Geometry showed something similar. In the Erlanger Program Klein had given a hierarchy of geometrical structures: topological, projective, affine and euclidean, in HF's terms from poor to rich. Children start by drawing irregular circles, anyone can see that. This was sufficient reason for Piaget to presume that their geometric development began with the topological structure. HF remarked condescendingly that those very same children were quite capable of distinguishing between correctly drawn circles and other figures. So what about this: the first modern mathematics course for the Dutch primary school should start with topological problems?[35] So HF's worries were justified.

3. LEARNING MATHEMATICS

> I have always known that my levels differed from those of the Van Hiele's and I have emphasized this on many occasions; my levels are relative rather than absolute, I would say, although of course Pierre van Hiele deserves all the merit for this discovery. But I have to admit that I have never felt the difference so clearly as I do now.
>
> From: Revisiting Mathematics Education

Van Hiele and his wife had concentrated on geometry education in their research. The subject matter, geometry as it filled school books in the fifties, was very suitable for thinking on different levels. Mind you, not levels in *subject matter*, but levels at which pupils *understood* that subject matter. For example children first learnt what a rhombus was by referring to it as a 'lozenge', the properties of which can be studied (discovered) with the help of all sorts of measuring and geometrical actions. The rhombus was an object of study at that time, therefore it was subject matter. At a given moment the pupils began to study the properties of the rhombus. This brought them to a higher level. The level where the rhombus is no longer the 'lozenge' of a short while ago, but for instance a quadrilateral with equal sides. The level theory has been explained hundreds of times in this way, for example to students of teacher training colleges.

However from the very start, HF lifted the level theory across the boundaries of geometry. In the first publications he already illustrates his version of the theory with the example of mathematical (complete) induction'. This had always been a tricky matter in the mathematics programmes of secondary education. A subject hard to teach and inaccessible for most pupils. HF indicated what caused this: people always began on far too high a level and did not give the pupil the opportunity to go through previous, lower levels of the learning process. At the same time he showed that the history of complete induction had not been exactly smooth either.

> For two millennia mathematicians had practised complete induction – with the ancient 'side and diagonal numbers' as the most profound example – when Pascal (and independently James Bernoulli) for the first time couched it in the now well-known principle; during this long historical period there isn't the slightest evidence for any conscious reflection on this practice, although figurate numbers would have been an excellent opportunity.[36]

HF did not refer to this piece of history in order to provide mathematics teachers who were slaving away in vain, with an excuse. There was more to it. He saw in the history of mankind rather interesting clues as to how educational learning processes should be arranged.

> (...) Start with figurate numbers, for instance square numbers as dot patterns. Look at the triangles below the diagonal. This is where you will find triangular numbers. How can you write down the fifth triangular number? And a few others? Do you see anything special, e.g. that it is always half of a certain rectangle? How would you describe the umpteenth triangular number?

$\frac{1}{2}n(n-1)$? Are you sure that this holds true for all triangular numbers? Now judge the formulae you have found on their merits. (...) And think about how you have investigated those formulae (...). etc.

This is how I imagine HF would lead a pupil via the (adapted) Socratic method[37] to higher levels, from the level of concrete dots and counting to the level of formulae and substitution, to the level of mathematical induction and proof, to the level of Peano's axioms and investigating systems such as the natural numbers, ...

Mathematics is an organizing activity. By thinking about organizing one enters a higher level in the learning process. This thinking is called reflection. Therefore continued learning requires reflection.

This section is about learning which has always been a central theme in HF's work, from his '*Didactics of arithmetic*' in 1942 up to and including his '*China lectures*' of 1991. Although the essence of learning has already been stated above (learning process and the raising of level through reflection) it might be worthwhile to follow the leitmotiv of HF's thinking in this field, merely by mentioning key notions and leaving a deeper going elaboration to others. Starting with his conjecture that 'mathematics does not teach to think' in the forties, to the understanding that he expressed recently by saying 'mathematics with the status of common sense' – a long road with various interesting stops in between.

Educating to think

In 1945 HF gave a lecture 'Educating to think' (not published) at Castle Rhederoord in De Steeg for members of the WVO. The core of his talk was that 'thinking' has so many more aspects than can be described with mathematics, and certainly more than can be described with mathematical logic.

Logical Thought

In 1951 a discussion was published about the question of whether mathematics education can contribute to the development of logical thought. This was still at the time of proof in geometry teaching. Mrs. Ehrenfest outlined in her contribution how she thought that basic concepts from the theory of numbers and of space (pre-eminently mathematical topics) initially developed on the basis of personal experience and thinking. She then showed how they developed further in the context of mathematics. Connections are found and possibly new basic concepts are added. So further development takes place on the basis of earlier experience and gaining new ones, but this time experiences with new connections and concepts.

She distanced herself from the old supposition that the logic of syllogisms could contribute to thinking. She could indicate good habits of thought: searching for the essence in a given situation, accounting for it, trying to formulate it clearly, confronting the answers that crop up with the entire complex of data that is already available and retaining an overall view. This is how logical thinkers do it. But to most people logical thought is not natural, so they have to be taught. Well, mathematics is an eminently suitable exercising ground to learn this. This is where she differed from HF who did not agree with this at all. He did not think that school mathematics (for secondary schools) was suitable for achieving something like that. Nor was the mathematics taught at school intended for this. The aim of 'learning to think' did not give mathematicians the liberty to ask for extra hours in the curriculum. The aim of mathematics education is to teach mathematics. Later when primary education had also inferred his perspective, he was to change his mind, but then with 'different' mathematics.

Learning Process

A few things have been said about this above: the possible links between what Mrs. Ehrenfest said about learning to think, what Mr. and Mrs. van Hiele put forward in 1957 and how HF interpreted this in his own way and added to it. It is clear that 'learning how to think' was no longer a subject of study. The focus was on learning processes and the discontinuities they contained.

Anti-Didactic Inversion

When the battle against New Math was at its height, HF wrote his first important educational book: *'Mathematics as an Educational Task'* (HF–1973). On page 103, in the chapter titled 'The Socratic Method', he summarizes his criticism of the new approach to mathematics education in two words: anti-didactic inversion. When didactising mathematics i.e. designing mathematics education to go with the subject matter given, one should not take the most recently composed mathematical structure as a starting point. That is a poor structure, whereas mathematicians in the past have finally come to that poor structure via rich structures and a lot of trial and error. They have had the chance to pass through a number of learning processes at various levels. Didacticians should start from the right side, so that their pupils can follow the 'natural' way. If you start from the wrong side, at the end product of mathematical activity, one commits an anti-didactic inversion.

Reflection

One does not learn mathematics in order to forget it again immediately, although New Math has certainly given that impression to many. You learn mathematics by doing. But this too is not enough if you do no more. Pupils should firstly know what they are doing, and what is even more important they should take the opportunity to think about what they and their fellow pupils have actually done. That is reflection during the process of learning. HF suddenly realized its importance while preparing a lecture for the conference 'Beweisen im Mathematikunterricht' at Klagenfurt.[39]

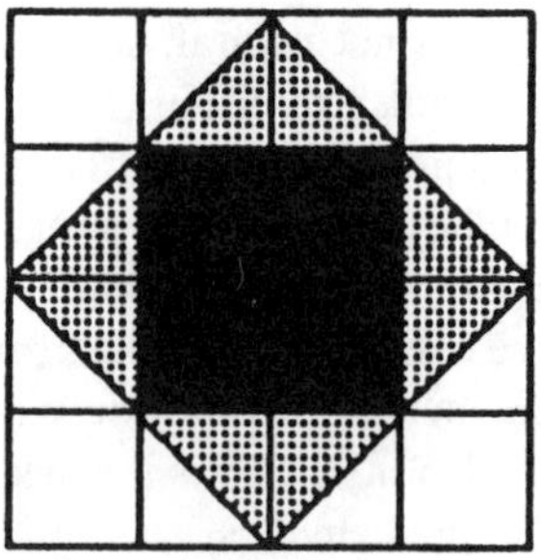

Fig. 3.

Mathematizing

What one has to do in order to learn mathematics, is to mathematize suitable subjects from reality, concrete situations with a problem that is either clear or not. This would be an organizing activity, roughly speaking, controlled by what occurs in the situation and supported by the mathematical means that are already available. In the course of time, HF added all kinds of variations to this which better and better expressed the specific aspect of mathematics as an activity.[40]

Re-invention

In order to obtain indications for the structuring of learning processes HF turned to history. How, in the course of time, had people discovered mathematics that today is the subject matter for schools? He devoted a chapter to this in HF–1973 that was mainly theoretical. The practical working-out follows later, in particular, when didactical phenomenology comes into view.

The Socratic Method

Socrates showed Menon that his slave had a lot of geometrical knowledge which merely had to be brought to the surface. In order to achieve this one had to ask the right questions at the right moment. HF had been inspired by these 'conversations' and (of course) interpreted them in his own way.[41]

This was clearly demonstrated on one occasion. HF was present during an instructive conversation between a student teacher and some primary school pupils. Accidentally the conversation also led to the same subject as the one Socrates discussed with Menon's slave, the duplication of the square. Socrates made the slave enlarge the square starting from the inside, by a method of trial and error.

HF interfered in the conversation and spread his handkerchief on the table. Then he asked them to fold it again in such a way that it would become half of the original handkerchief. So halving instead of doubling, from the outside towards the inside, with the material in view from the start and using material that was easy to fold. The pupil needed no help to fold the corners to the centre (Figure 3).

The Socratic method then to help pupils with re-inventing and reflection to follow, which HF called 'guided re-invention'.

Development

When writing HF–1973, HF grasped the opportunity to observe younger children. in '*Walks with Bastiaan*', he mainly observed spontaneous developments and he would wonder at the sudden insight, 'Aha-Erlebnisse', that often went hand in hand with violent emotions. When Bastiaan had been given a chemistry set by a friend, grandfather and grandson together started to search for Archimedes' Principle. Now HF had to interrupt more often and to 'teach'. One inference was that an adult can stimulate the cognitive development of a child by means of good questions and suitable material.

Rich Contexts

At the Dr. William Drees School at Arnhem, the design school of Wiskobas, designers put much old and new subject matter in situations that were familiar to children. For example open sentences in the context of getting in and out of a bus, ratios in the context of Gulliver's Travels, counting large numbers in the context of picking flowers, measuring area in the context of allotments etc. HF was often present in class and saw how that approach inspired children to interaction and so stimulated reflection (about one's actions and those of others).

Pupils and Productions

Sometimes designers invent things too. Allowing children to make their 'own productions' is one of them. A didactical invention! The children of grade four for instance are asked to make a book about arithmetic for the children in grade three. Or pupils from grade eight are allowed to write a manual for a pocket calculator, to be used by pupils who can not yet work with it. In order to carry out such assignments, children have to think about their own learning process. How did I do that? What did I find difficult? How did I find out? Did I understand it completely then? etc. Thinking about one's own thoughts, reflection therefore.

Actually the term 'own productions' was not invented by HF himself. Although he came very close to this didactic phenomenon in his analyses of educational learning processes, this was invented by other Wiskobassers[43] who experienced its tremendous impact in class.

Common Sense

HF put the above in a new light in three masterly articles entitled 'Mathematics phenomenologically'. What is new about it? Well the story moves between two poles: mathematics as something 'from outside' and common sense as something 'from inside'. In order to know something difficult for certain, man can make use of mathematics (in case of need one can ask a mathematician). Simpler matters he can solve for himself with his common sense. In HF mathematics had become an integrated part of his common sense.

HF as a learner can be the model of what has so far been said about learning. For we assume that HF used to learn like this: constructing, reflecting and, particularly, under his own steam. When something was worthwhile it was simply remembered. He did not have to pay special attention to drumming in subject matter. If this hypothesis is correct we may have found an explanation for the fact that HF never did anything about the problems of poor calculators at school.

This was done by two other leading mathematicians, Peter Hilton[44] and Hassler Whitney[45]. They started from the prevailing education in classrooms and analysed the difficulties encountered by the children. Thus they produced some interesting ideas and came up with practical didactical inventions. In recent developments which are also due to the discussion between designers, remedial teachers and ortho-didacticians, the work of Hilton and Whitney can easily be recognized.[46] With this material, less talented pupils can also acquire elementary arithmetical skills into their own ability.

But let us return to HF. As I wrote, he did not need any support to memorize mathematics. And this did not merely apply to mathematics but to anything he considered worth remembering. This showed in the fact that he was able to share his knowledge with others at all levels. New insights and practical information were immediately incorporated in the larger whole of knowledge, of his sharpened common sense. All his work in education illustrates this.

It is therefore not surprising that HF saw learning mathematics ultimately as a development process of common sense that acquires content and direction by looking for certainties in realistic situations.

> Certainty should be aspired to and the way this is done characterizes mathematics as an activity that time and again induces improved versions of common sense.

Via this line of approach all that was said earlier can be given a place. Please consider this pattern of a theory:

We are not merely concerned with *thinking* as such, but the application of *common sense*. By the activity of *mathematizing in realistic situations, interactively* with others and *reflectively*, common sense is continually brought to a higher state. This implies that it can be applied more intensely in more situations. Initially one has a lot of 'natural' common sense but when challenged by rich contexts, inspired by the opportunity to make inventions oneself and guided by someone who knows about mathematics when looking for more certainty, then common sense will be *enriched with mathematics*. The essential means to integration of common sense and mathematics, or if you like to assimilation of the latter into the former, is *reflection*. Mind you, mathematizing can very well take place in a *mathematical context*. For mathematicians also use their common sense in particular when working on the boundaries of their science.

4. EDUCATIONAL DEVELOPMENT AND DESIGN

On 26 January 1971, IOWO, the Institute for the Development of Mathematics Instruction was established as the executive arm of the governmental Commission on Modernizing the Mathematics Curriculum. The end of a five year period of activities invites a review.
IOWO is not a research institute; its members do not regard themselves as researchers but as producers of instruction, as engineers in the educational field, curriculum developers.
Engineering needs background research and can produce research as fall out. Though both of them will be visible in the present account, its nucleus is our productive work, represented by a few specimens, and embodies our view on mathematics as a human activity and on curriculum development as a classroom activity, guided by curriculum developers, in a continuous contact with all those interested in mathematics education. IOWOers are people rather than an abstract entity but their work is a collective achievement. So is the present volume, written by individuals and edited by a team. All our national publications are meant to invite response from the field. I hope the present call will be heard abroad.

Hans Freudenthal

This was the Preface to *Five Years IOWO*, a special edition of *Educational Studies in Mathematics*, published in August 1976 on the occasion of HF's retirement as director of the IOWO. The insights in educational development (which is more than mere curriculum development), gained in the previous five years were clearly visible to insiders: designers as engineers, research in aid of development work, designing material for mathematical activities based on a vision of mathematics education, presence in the classroom, together with children and teachers, material and education were creatively prepared and summed up critically in the IOWO team, often in dialogue with interested parties from outside the IOWO. The subtitle of the special (edition) was: *IOWO Snapshots*. The first 125 pages consist of teaching materials. First attention is given to geometry (a show-box with pre-school children, an imaginary island for young primary school children, Ship Ahoy – a theme for the middle groups and mathematical world orientation for the oldest primary school pupils). Then it is focused on proportions (e.g. a theme about Gulliver's Travels, making enlargements with the help of an overhead projector in the first form, Look and Luck, and research into the ratio table). HF had contributed to the teaching materials which have been shown. The development groups for primary and secondary education held weekly meetings where they would design material together and exchange experiences they had had in class. HF was always present, he listened to the stories from daily practice, directed the first ideas and inspired at crucial moments. He also came to the design school in Arnhem every week and observed the learning processes, he would speak with

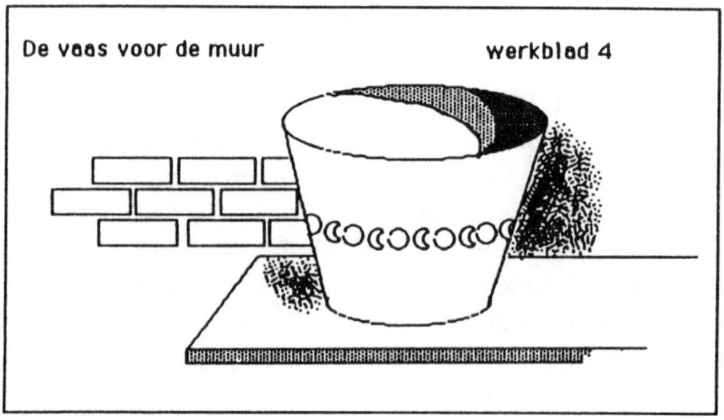

Fig. 4. .

the children and teachers and frequently he would come up with ideas that the designers could develop further.

The Vase

In Wiskobas Bulletin number 2, volume 3, there is a picture of a vase on page 156 'worksheet 4' (Figure 4):

This is a worksheet based on an idea of HF. He got this idea during observations in the first form when all sorts of objects were counted 'in one's own surroundings'. The vase was on a cupboard at the back of the class. HF saw at once the possibilities it had to offer. Is it possible to count something you don't see? In the case of the suns and the moons on the vase this is indeed possible. But you have to be clever then, to organize. And if you look carefully you can see a good structure. Is there anything special as far as you can see?

The lectures given by HF during his IOWO period, were always full of examples like that of the vase. But they did not stand alone, they were paradigms in an educational-theoretic framework, which was now being fully developed.

Major Problems

In 1980 the fourth ICME conference was held, this time in Berkeley. HF had been personally invited by Pollak to deliver the opening address. The assignment said: deal with the major problems which will face mathematics education in the coming years (the eighties). MAJOR PROBLEMS OF MATHEMATICS EDUCATION.[48] The first nine 'major problems' varied from *'Why can Jennifer not do arithmetic?' to 'How to create suitable contexts in order to teach mathematizing?'* HF was mainly drawing attention to 'learning' and 'development' in California at the time, in other words creating learning processes. Educational problems are not like mathematical problems, HF told his audience. The latter can 'simply' be solved, the former have to be worked out.

Is this not a marvellous display of problems of mathematics education? Show me the wonderful world where these problems might be solved, would be solved! Rather than in the armchair or in the laboratory, educational problems are solved in the educational process (...)

Didactical Phenomenology

But vision and a few creative teachers are not enough in the end. HF was aware of this. He remained active in the team and at school, but behind the scenes a new instrument was created for educational development. It advanced the knowledge of mathematics, its applications, history, the well thought-out observation of learning processes, didactical notions in development, analyses of school books and much more. As I wrote earlier, the staff of the IOWO frequently found a chapter in their pigeonhole.

What was it all about? So far curriculum developers had occupied themselves with subject matter (at program level) and with textbook authors (at classroom level). Choosing subject matter, arranging it, concretizing, visualization, testing, HF did not want to discuss all that.[50] What he cared about was mathematizing. The question was: what is worth mathematizing, and how would we, didacticians, like learning processes to proceed in those cases? Well, HF found a solution to this problem. It was great and a lot of work. In order to get the idea one needed a change of perspective. His thoughts were like this: mathematics is a cognitive instrument (public knowledge) to organize, structure and mathematize parts of reality. Through this organizing, structuring and mathematizing an individual makes mathematics his personal property. So go and look for those phenomena in the surroundings of children which can be mathematized by certain parts of mathematics. Go back in history in particular and see how it was discovered by trial and error at the time. Describe how mathematics organizes the phenomena etc. Show how the 'thought-thing' (noomenon) describes the 'phaenomenon' (phenomenon), analyses it and makes it accessible for calculation and thinking activity, helps one to understand it better, allows one to predict etc. This is called 'phenomenology', the phenomenological method as HF saw it. He published in 1971 a 'Phenomenology of the number concept'[51] and he would refer to it in numerous publications afterwards. Later he was to modify it, and what crisscrossed in 1971, the mathematical phenomenological point of view and the didactic perspective, was then unravelled. With a *phenomenological analysis* it was merely a question of a cognitive achievement which expresses how the thought-thing can organize a phenomenon. *Didactical phenomenology* is directed in particular at the acquisition of all that which is needed for organizing. A large number of phenomena passes in review: length, natural number, real number, fraction, ratio, quantity, set, angle, variable, function, ... Are these thought-things or phenomena, one could ask. Do not panic, for when one has reached a higher level in the learning process, mathematics which one acquired earlier close to daily reality, becomes itself a phenomenon to be organized.

The above reflects what HF published in 1971. Later he elaborated in various places and for different readers the didactical (phenomenology) of natural number. Most elaborate in HF–1984, it begins with a phenomenology (the mathematical-historic entrance) and subsequently he tested this didactic-phenomenologically (the psychological-didactic entrance).

An interesting novelty here was HF's changed point of view on the phenomenon of 'conservation'. This invention by Piaget had earlier been described by him as 'artificially created'. But now he considered it seriously and even gave it a place in his phenomenological analyses.[52] This is only a minor matter; what is more important is the fact that HF's didactical-phenomenological view of natural number led to a large number of leads for designers of initial arithmetical education. The same applies to 'measuring', 'ratios' and fractions, just to mention a few from which designers have already benefited. Whoever has the opportunity to consult primary mathematics textbooks, will in essential places recognize HF's preliminary work – mathematical-didactical analyses which are unique in the world.

At the same time HF created a new paradigm by purposely integrating so much knowledge of varius fields in aid of educational development. For didactic phenomenology clearly shows how HF's theory, the pattern of which was described at the end of the previous section, worked in the case of HF. What was called 'mathematizing' in the previous section, should here be replaced by 'didactizing', horizontally, through the observation of learning processes, and vertically working on the formation of theory.

Forwards

The word processor informs me that you have read 13 452 words so far, not counting references. Of course this does not cover hundred percent HF's work on (mathematics) education development. I dare not say to what extent. Whoever wants to know more, is referred to HF's work itself. He or she will have sufficient material to last him/her a long time.

Those who want to go on from here on the basis of Freudenthal's work can do so and this is already happening. At the academic level a theory of realistic mathematics education is being worked out. Future primary teachers are trained at colleges in the manner of Freudenthal. At primary schools realistic methods are becoming increasingly popular. In the upper grades of secondary education the new programmes have the same signature. The lower grades are yet to come. People are working on this: there is a large team of designers in the framework of the Van der Blij Committee, now C.O.W. (Committee for Mathematics Education Development). Freudenthal was an advisory member of this committee. If we can assume that they have listened to him we will find at all levels the result of HF's work on the development of mathematics education.

And I am convinced that Freudenthal would have added that this was possible only because he also listened to them, for educational development is a matter of teamwork.

NOTES

[1] I owe many thanks to my colleagues of the library and the BSO department of SLO, who managed to convert my quest into concrete products.

[2] But this was not easy to do. HF had to make a serious effort. In his introduction to Make a note of that Hans (p. 11): 'Every well-turned phrase I write is considered, weighed and turned around until I have achieved the ultimate level of spontaneity. Only then is it written down and printed out. I want it to look as if I have jotted down what just occurred to me.' This did not remain unnoticed, for on December 12 1984 the minister for educatoin handed to Prof. H. Freudenthal (79 years old) the Golden Quill, which is the cultural prize of the Royal Dutch Publishers league (KNU).

[3] *Make a note of that Hans*, p. 355 (in Dutch).

[4] I want to express my gratitude to my colleagues of the Freudenthal Institute for enabling me to see the manuscript. While reading it I often thought: 'Why on earth should I want to put in an article what HF wrote in a book so well and so clearly?'

[5] See Hans Freudenthal and Fred Goffree, *Mathematics Education. Ten years of curriculum development 1975–1985*, SLO Enschede 1987 (in Dutch). Chapter 2 is called 'prehistory'.

[6] Ibid, p. 22.

[7] Study group for the Renovation of Didactics and Education (WVO), the Dutch department of the New Education Fellowship, in which Kees Boeke and Suus Freudenthal–Lutter were also involved.

[8] Mr. Metselaars of the National Archive kindly guided me through the material that had only recently been acquired.

[9] *Make a note of that Hans*, p. 337.

[10] Ibid, p. 178 and further.

[11] As far as I could see this manuscript was not published either. It is also in the National Archive for Noord-Holland at Haarlem.

[12] For example in *Mathematics as an Educational Task* from 1973, p. 119.

[13] Can mathematics education contribute to the education of thinking? Discussion between T. Ehrenfest-Afanassjewa and Prof. Dr. H. Freudenthal, publication of the Mathematics Study group of the Study group for Renovation of Didactics and Education, Muusses, Purmerend MCMLI (in Dutch).

[14] In '*I used to be a mathematics teacher*' (SLO, Enschede 1985, in Dutch) Pierre van Hiele tells something more about it. This booklet also contains interviews with other mathematics teachers. Three are in the photograph on p. 26 together with HF. The interviews were dedicated to HF on the occasion of HF'80.

[15] See P. M. van Hiele, *De Problematiek van het Inzicht* (The Problematic Nature of Insight), Utrecht 1957.

[16] Between 16 March 1946 ('Calculators are winning the war') and 18 August 1962 ('The end of the construction industry holiday') 89 articles appeared. For those who would like to see all titles, they should take '*The anniversary book. Prof. Dr. Hans Freudenthal 70 years old*' published by the Institute for Mathematics at Utrecht, 17 September 1975 (in Dutch) and look on page 136.

[17] The CMLW had been instituted on 12 June 1961 and it was officially installed on 19 July by the Secretary of State Stubenrouch. For further information on CMLW see e.g. 'Ten years of curriculum development', p; 24, 25.

[18] '*Make a note of that Hans*', p. 350: '(...) Could you have stopped the flow of New Math which was to rage on like a bandjir thanks to Royaumont? Well, it need not have become a bandjir. (...)' Also '*Ten years of curriculum development*', p. 24.

[19] In 'Education into thinking' of 1945, a similar summing up of the situation occurs. But then it was in order to illustrate the necessity of a new attention to word sums and transfer of training.

[20] Freudenthal, H., Why to teach mathematics, so as to be useful, *Educational Studies in Mathematics*, Volume 1, nos. 1/2, 1968, p. 6.

[21] As Mrs. Krygowska from Poland, H. O. Pollak (Bell Company, US), André Révuz (F), Willy Servais (Belgium), H. Behnke (FRG), T. J. Fletcher (UK), Hans-Georg Steiner (Bielefeld), Arthur Engel (Stuttgart) and Maurice Glaymann (Lyon).

22 1973 *Mathematics as an educational task.* (HF–1973)
1978 *Weeding and Sowing. Preface to a science of mathematical education.* (HF–1978)
1983 *Didactical phenomenology of mathematical structures.* (HF–1983)
1990 *Revisiting mathematics education. China Lectures.* (HF–1991)

23 Published by Van Walraven (in Dutch), Apeldoorn 1984.

24 As has been said before, HF distinguished between 'concept of number' and 'number concept'. He did not think that all theories of concept attainment applied to learning mathematics, as he saw it. Instead he preferred to say 'the constitution of mental objects'. But in the present case it is a question of didactic notions. Is this allowed? Yes, for in his last book HF draws a parallel between 'mathematizing' (as an activity to re-invent mathematics afresh) and 'didactising' (as activities that aim to make others re-invent afresh). This justifies the metaphorical sense here.

25 HF–1991, p. 147 (of the manuscript).

26 Freudenthal, H.: Fiabilité, validité et pertinence-critères de la recherche sur l'enseignement de la mathématique, *Educational Studies in Mathematics* 13 (1982) 395–408.

27 HF–1978, p. 121: '(...) Properly speaking it is even more sophisticated. It is not only the diagnosing teacher who gets blindfolded. It is the so-called double-blind method. The eyes of the designer of the diagnostic test are also bandaged by means of educational theory. It is compulsory that he understands nothing of the didactic intentions of the course or textbook for which he designs a battery of diagnostic tests; he is bound indeed to investigate whether certain 'objective' instructional aims are attained – such as solving quadratic equations – rather than whether the learning process intended by the author of the course took place (...)'.

28 In 1979 and 1980 HF gave two celebratory lectures explaining inter alia the essence of development research. The first took place on 19 March 1979 on the occasion of 10 years of Wiskobas: 'Filling in–fulfilling' (in Dutch) and then as a contribution to 15 years of Didactics in Utrecht 'A view on being educationally active' (25 January 1980) (in Dutch).

29 'Research in Education', in *Pedamorfose* XLV 57–66.

30 See for instance the inaugural lecture of Prof. Dr. Jan de Lange: *Heart versus hard. Renovation of mathematics education*, Utrecht, 16 January 1991.

31 In a reaction to a report on Mastery Learning, published by the KPC (under the name of K. P. M. J. Hollman, not dated), HF wrote a piece: 'A flaw in Mastery Learning'. This probably referred to a part of the so-called Mavo-project, in which the subject matter for mathematics had been organized afresh, in accordance with the model of Mastery Learning. In this article HF quotes the following explanatory note that belongs to this model: 'Mastery Learning wants to achieve the situation that (almost) all pupils attain the minimum targets of education. For this purpose classical education that is uniformly given to permanent groups, is frequently interrupted to hold a diagnostic test. By means of such a test, which systematically evaluates all the targets of a lesson unit, the teacher can come to a diagnosis. He can accurately ascertain at which point the pupils have not yet attained the targets. Then he is able to indicate a (possibly alternative) individual route of learning, enabling the pupil to master the entire subject matter after all. As soon as these deficiencies have been remedied, he continues with the next part of the subject matter with the whole class.'

32 Hans Freudenthal, Ways to report on empirical research in education, in *Educational Studies in Mathematics* 10 (1979) 275–303.

33 Column: Recording empirical research in education on the homeground (in Dutch), *Pedagogische Studiën* 1981 (58) 141–143.

34 On page 91 (HF–1978) follows one of the conclusions which the reader may meanwhile have expected: 'whether something is knowledge or comprehension does not depend on its mathematical content but on its place in learning process, or at least on its instructional context.'

35 This was 'Thinking and arithmetic. The new approach to arithmetic at primary schools' (in Dutch), published by Malmberg. It was a translation of the work of Nicole Picard of the Institut Pédagogique National: 'A la conquête du nombre. Topologie'. Published by J. van In & Cie S.A.

36 HF–1991, 2.3.4.

37 Much better dealt with in HF–1973, p. 99, than here.

38 HF refers here to problem sums that occupied an important place in arithmetic education at the

time: word sums subdivided into certain groups each with a specific way to solve them. Kohnstamm had worked on this in the thirties in the framework of the psychology of thought by Selz. The idea of 'general ways of solving' for thinking, resulted in arithmetic education in the opposite, the teaching of algorithms.

39 Konstruieren, Reflektieren, Beweisen in phänomenologischer Sicht, Schriftenreihe Didaktik der Mathematik, Klagenfurt 2 (1979), 183–200.

40 See for instance Panama-Post, volume 8, no. 4, 1990. Here HF published the third and last article in a series under the title 'Mathematics phenomenologically'. The three articles can be considered as a pre-publication of HF–1991.

41 HF–1973, p. 100: 'I like such lectures, such lessons. One person is speaking, but it is not just a one-man-band, the other party is heard too. When he has finished, everybody has decided on his own what is right. The lecturer (let us be grateful) was only our midwife – and this is what Socrates pretended to be. He has delivered us of our own thoughts rather than his.' That is the Socratic method or dialectics, or as I would prefer to say, the method of the thought-experiment.

42 See 'Bastiaan's Lab', in *Pedamorfose* XXX, 35–54 or 'Bastiaan's experiments on Archimedes' principle in *Educational Studies in Mathematics* 8 (1977), 3–16.

43 Jan van den Brink toyed with this idea in the starting phase of Wiskobas and Leen Streefland paid a lot of attention to this in his thesis. See Brink, J.v.d., *Realistic arithmetic instruction for young children* (in Dutch), OW&OC, Utrecht 1989 and Streefland, L., *Fractions in Realistic Mathematics Education. A Paradigm of Developmental Research*, Kluwer Academic Publishers, 1991.

44 See for instance Hilton, P. and J. Pedersen: *Fear no more. An adult approach to mathematics*, Addison-Wesley, Menlon Park 1983.

45 See for instance Whitney, H.: Taking Responsibility in School Mathematics Education, PME 1985.

46 For insiders I can, as a case in point, use terms such as 'five structure', 'empty numberline', 'calculate up to one hundred' as a basis for 'mental arithmetic' and 'estimating arithmetic'.

47 See HF–1978, p. 176: The theoretician in the team. '(...) First of all, however, the theoretician should be the conscience of the team by watching its goals even if they have not been made explicit and were only formulated as a result of watchfulness and as a means of warning. In particular I mean formal goals which express a backgrounds philosophy (...)'.

48 Published in Educational Studies in Mathematics 12 (1981), 133–150.

49 *Didactical phenomenology of mathematical structures (part 1)*, OW&OC, 1984, p. 39.

50 In 'Major problems' he began as follows sub problem 8: 'Perhaps you will complain that up to now I have paid almost no attention to subject matter and its didactics. If by subject matter you mean some chapter of a textbook, you will be disappointed. These are not major problems. But I agree that teaching is always teaching something. Something rather than anything. Something worth being taught. But what is worth being taught?'

51 In *Philosophical perspective on society and science*, pp. 225–235.

52 Conservation is what mathematicians would call 'invariance', invariance under a certain transformation. HF believed that these transformations should be investigated. In the case of Piaget's most famous test, with a row of sweets below which another row of sweets is put, just as many but further apart – with the question: which is more? – a child is lured into a trap. For the transformation causes more room to be taken up in any case, because the bottom row is longer.

I owe many thanks to my friends Alan Bishop and Geoffrey Howson from England, and Thomas Cooney from the United States for helping me to tell HF's story to an English speaking audience.

National Institute for Curriculum Development SLO
P.B. 2041, 7500 CA Enschede, The Netherlands
University of Amsterdam

HENK J. M. BOS

'THE BOND WITH REALITY IS CUT' – FREUDENTHAL ON THE FOUNDATIONS OF GEOMETRY AROUND 1900

ABSTRACT. In the late 1950's Freudenthal published several articles on the history of geometry around 1900, in particular on Hilbert's innovative approach to the foundations of geometry. Especially his essay-review of the eighth edition of Hilbert's *Grundlagen der Geometrie* has become a standard reference in historical studies of geometry. In the present article I discuss Freudenthal's contribution to our understanding of the development of geometry at the turn of the century and his style in historical research.

In 1957 Freudenthal published in the *Nieuw Archief voor Wiskunde* what present-day editors call an essay-review of the 8th edition (1956) of Hilbert's *Grundlagen der Geometrie*.[1] In a number of later articles[2] he returned to the history of the foundations of geometry, working out items which in the review were only briefly treated or following some other lines into the history of nineteenth-century geometry. The *Nieuw Archief* article has become a standard reference in the literature on the history of nineteenth- and twentieth- century geometry. The Freudenthal memorial issue of *Educational Studies in Mathematics* was a welcome occasion for me to reread the article and to consider its results, its style and its reception.

Hilbert's *Grundlagen der Geometrie* appeared in 1899. It contained a complete axiom system for Euclidean geometry. Euclid's axioms were far from complete, and hence traditional geometrical argument was a mixture of logical deduction and (usually tacit) reference to the perception of spatial forms. Hilbert's book established the purely logical approach to the study of geometry with special interest in the interdependence and consistency of the various axioms and groups of axioms. Hilbert analysed these questions with great thoroughness, testing for consistency and interdependence by providing models of different geometries which verified various combinations of axioms. The book inaugurated, in the words of Schmidt "a new type of axiomatic thinking"[3] eagerly taken over by many mathematicians and applied with great success in a wide field of geometrical research in the twentieth century.

In his review Freudenthal discussed the main parts of Hilbert's book in considerable detail; he dealt with the development of these parts in the subsequent editions and with the contributions of others to lines of enquiry opened or suggested by Hilbert. The main themes he covered in this manner were: the choice of the axioms, the models which Hilbert used to analyse their interdependence, his

Educational Studies in Mathematics **25:** 51–58, 1993.

theory of proportions, his calculus of line segments, and the theory of constructions in relation to the algebraic structures underlying the geometries.

These are technical issues and the present article is not the place to review them. Rather I shall concentrate on Freudenthal's more general remarks and on his overall evaluation of Hilbert's work.

Freudenthal showed that much of what Hilbert presented in his book, both as to techniques and as to programme, had been initiated by others in the 1880's and 90's. Thus[4] Pasch in 1882[5] had started the strictly logical study of geometry, requiring that all deductions were based exclusively on the rules of logic and on the relations codified in the axioms. Also the Italian geometers, Peano, Padoa, Pieri and others, had focused on the strictly logical study of geometry. The use of implicit definitions (Hilbert's axioms defined points, lines etc. implicitly by giving the rules they obeyed) went back to Gergonne. Counter examples had been used to prove independence of axioms in non-Euclidean geometry; Peano and Padoa had improved that method. Von Staudt had created a calculus of line segments. Wiener had studied the role of incidence theorems such as those of Pappus and Desargues. Even the term "Grundlagen der Geometrie" had been used before by Riemann, Lie and Killing.

The contributions of Pasch and the Italian geometers inaugurated the strictly logical study of geometry, requiring that the geometrical theorems and proofs were valid irrespectively of the meaning of the terms used, and a fortiori independent of the perception of geometrical objects. Hilbert, Freudenthal writes[6] is often seen as the originator of this approach, of which indeed he coined the slogan with his famous dictum that the theorems of geometry should remain valid if the terms point, line and plane were replaced by table, chair and beer mug.[7] However, Pasch and several Italian geometers expressed this requirement earlier, formulated it as a program for geometrical research and initiated studies from that point of view. Nevertheless Freudenthal does claim that it was Hilbert's *Grundlagen* that essentially constituted the[8] "Wende zur Axiomatik"[9] in mathematics. What, then, *was* Hilbert's unique contribution? Partly it was the comprehensiveness of his work. The "tiefgründig durchexerzierte Beispiel einer Axiomatik", as Freudenthal characterized the *Grundlagen*, carried much more conviction than incidental studies and programmatic statements.[10] But the aspect of the book which Freudenthal saw as the most crucial one in the turn towards axiomatics was Hilbert's attitude to contemporary philosophical issues related to geometry.

At the end of the nineteenth century the interest in the foundations of geometry was far from being merely a mathematical concern, it related to major philosophical questions about the nature of space and of our knowledge of it. Kantian philosophy, the appearance of non-Euclidean geometry and the prestige of mathematics and natural science combined to make these questions, which unexpectedly appeared not to admit easy and direct answers, urgent indeed. Pasch accepted this urgency; in Freudenthal's words he was anxiously concerned[11] to derive or at least to justify the axioms of geometry from experience. Despite his insistence on purely logical reasoning, Pasch saw geometry as the science of real space. And all[12] other geometers who, before Hilbert, advocated strict logical axiomatic reasoning in geometry, likewise saw geometry as the science of real space and considered logic and rigorously deductive argument as safeguards by which the epistemological and ontological problems were located at the choice of the axioms and could be discussed, if not settled, exclusively there.

According to Freudenthal, Hilbert's unique and principal merit lay in his different attitude to these philosophical questions: he simply disregarded them. For Hilbert, unlike his predecessors, the study of the foundations of geometry was the study of the logical relations between axiom systems, irrespectively of the question whether or not these axioms adequately or optimally described the space or the space objects we experience or intuit. There are no philosophical arguments to justify the axioms in the *Grundlagen*, Hilbert begins abruptly:

> Wir denken uns drei verschiedene Systeme von Dingen: die Dinge des ersten Systems nennen wir *Punkte*...

and Freudenthal comments:

> damit ist die Nabelschnur zwischen Realität und Geometrie durchgeschnitten. Die Geometrie ist reine Mathematik geworden, und die Frage, ob und wie sie auf die Wirklichkeit angewandt werden kann, beantwortet sich bei ihr ganz wie bei irgendeinem anderen Zweige der Mathematik. Die Axiome sind nicht mehr evidente Wahrheiten, ja es hat nicht einmal mehr Sinn, nach ihren Wahrzeit zu fragen[13].

Freudenthal argues at length how novel this attitude was. He also reports his own experience, while studying the sources, of how easily one reads the now so self evident Hilbertian ideas into texts which do not contain them at all. He illustrated this[14] by the case of Poincaré's *La Science et l'Hypothèse* of 1903. From earlier reading of that book Freudenthal had concluded that Poincaré had been influenced by Hilbert. However, retracing Poincaré arguments to his earlier writings in the course of a discussion among French mathematicians and philosophers in the early 1890's, and doubting that Poincaré could actually have anticipated Hilbert, he reread more carefully and arrived at the contrary conclusion: Poincaré and his interlocutors were convinced that geometry should fathom real space; abstractness in their view was not logical abstractness but merely the requirement that, for instance, arguments about triangles should be independent of the particular shape of the triangle one draws in illustration. Freudenthal comments:

> Es ist wirklich erstaunlich, wie man so ganz andere Auffassungen *nach* Hilbert's "Grundlagen" zwanglos im Hilbertschen Sinne interpretieren kann. Die menschliche Sprache ist offenbar ein merkwürdiges Instrument. Der Historiker nehme sich in acht: ein ganzes Buch kann in einem anderen historischen Kontext etwas ganz Anderes bedeuten – Science et Hypothèse ist ein Beispiel.[15]

An important observation indeed, and the phenomenon well illustrates how novel Hilbert's contribution was and, at the same time, how difficult it is for later generations to re-experience that novelty.

Freudenthal was not the first to survey the history of the foundations of geometry around 1900. Dehn had dealt with the historical development of the field in a lengthy appendix to the second edition of 1926 of Pasch' *Vorlesungen über neuere Geometrie*;[16] Schmidt had analysed Hilbert's contribution to the field in a note in the second volume (1933) of Hilbert's *Gesammelte Arbeiten*;[17] Nagel had written on the subject from an interest in the modern developments of logic in an article in *Osiris*[18] and Weyl had devoted several pages to Hilbert's axiomatics in a long overview of Hilbert's mathematical work published in the *Bulletin of the American Mathematical Society* shortly after Hilbert's death.[19] Indeed, much of the factual information in Freundenthal's articles had been recorded before, although he was more informative about the contributions of the Italian protagonists than was usual.

Yet, reading Freudenthal's articles and comparing them with these earlier works one finds that he added something which was not yet there in the earlier writings. Indeed one may say that Freudenthal's article began the truly historical study of the episode. It marked the transition from chronological reports and explanations of mathematical achievements, with Hilbert's *Grundlagen* as the focus through which most lines of development passed, to a view of Hilbert's achievement in its contemporary context, written with distance and nuance, notwithstanding an evident personal involvement with the subject. Freudenthal re-drew the picture of the foundations of geometry at the turn of the century and made it historical. He did not diminish Hilbert's merit, on the contrary, but he made his achievements understandable, compared them with those of his contemporaries, and located their essential novelties.

Despite the new historical dimension, however, Freudenthal felt close enough to Hilbert and his contemporaries to incorporate considerable mathematical criticism in his article. He indicated where he considered Hilbert's approach cumbersome, unnecessarily complicated, guided by interests now no longer accepted, or simply wrong.[20] He explained where modern terminology had become better, deploring that certain Hilbertian expressions were still used out of misplaced reverence for the master. He was not a dispassionate analyst of past mathematics. He used strong language on occasion, not so much about Hilbert, but some others were decidedly out of Freundenthal's favour. One was Klein, whom he blamed repeatedly for failing to grasp the logical function of a model.[21] Another was Russell, whose *Essay on the Foundations of Geometry* of 1897 he characterized as "Komische Note", because Russell was unaware of the discussions on the subject in Germany and Italy since c. 1870.[22] Here wrote Freudenthal the mathematician, very much his own irreverent self, and not afraid to shock contemporaries. Some were duly shocked. Bottema protested in a reaction criticising several factual statements by Freudenthal, mostly concerning Klein. However, one feels that it was primarily the irreverence that had irritated Bottema; significantly he contrasted Klein's criticism of Helmholtz, expressed with due "Ehrfurcht", with Freudenthal's criticism of Klein, in which the Ehrfurcht was sadly lacking.[23]

For all the historical insight in the articles, it seems that Freudenthal, at least in 1957, did not like to be considered a historian of mathematics. At the outset of one of his articles on the subject (1960) he more or less excuses himself for turning to the history of his science and reports the impression that much historical writing has to be characterized as "die Menge desjenigen, das jene Verfasser sich gegenseitig aus den Fingern gesaugt haben".[24] He wrote indeed as a mathematician; his view of the development of mathematics was one of doubtless progression; he was not afraid to judge the achievements of earlier mathematicians and he based these judgements primarily on two criteria: the mathematical quality of their arguments and their readiness to take up the modern progressive views in mathematics.

This approach certainly brings liveliness and personality to the study of the past, but I feel also that something is missing. For can it really be that the interest in thinking and writing about the past lies primarily in deciding who was the cleverer mathematician and who was more quick to jump on the progressive band wagon? It is not possible to go beyond judgement to explanations or questions? It is, and later historical studies have gone beyond the restrictions implied in Freudenthal's approach. I shall briefly mention two examples concerning the two mathematicians who did not pass Freudenthal's test: Russell and Klein. In her studies of nineteenth-century geometry in its cultural and philosophical context[25] Joan Richards has shown how strongly the attitudes and convictions of mathematicians about the relation of geometrical theory to real space were connected with philosophical and ideological positions which in turn related to the different and fluctuating social forces on whose support the mathematical community depended. In that context Freudenthal's irreverent characterisation of Russell's *Essay on the Foundations of Geometry* as "Komische Note" becomes irrelevant, and the significance of the *Essay* can be convincingly determined; it marked the end of the hope for a "unitary truth structure"[26] for geometry and advocated a definite split between the educational and psychological aspect of space conceptions on the one hand and the logical and mathematical aspects on the other.

Klein's conception of geometry, and of mathematics in general, has recently attracted much interest from historians of mathematics who see him as a central and crucial figure in the institutional and social history of mathematics in the Wilhelminian and Weimar periods in Germany. That interest, which arose in the wake of influential social studies of science in the same period[27] was further inspired by a wish to understand what went on in mathematics and science in Germany before the dramatic and drastic developments during the Nazi period. It appears that Klein and Hilbert, and their different approaches to foundations of mathematics, fit into a wider scheme of developments, which, as Herbert Mehrtens argues in a recent book devoted precisely to these issues, can be usefully studied in a dualistic scheme of modernist versus counter modernist movements in all spheres of public, cultural and political life.[28]

The fact that later research has found new questions and directions does not detract from the value of Freudenthal's articles on the foundations of geometry at the turn of the century. They are informative, lively and occasionally provocative, and what more would one want? For me personally they have an additional charm. I read them for the first time in the early 1970's in connection with a seminar by Freudenthal on the history of nineteenth-century geometry in which, as a young member of staff of the Utrecht Mathematical Institute, I participated as an assistant. The liveliness of the articles reminds me with pleasure of the same quality Freudenthal brought into the sessions of the seminar.

NOTES

1 Freudenthal (1957c), review of (Hilbert 1956); Freudenthal had incorporated historical remarks on nineteenth-century geometry in his earlier studies on the foundations of geometry, cf. his lectures (1957a, 1957b) both delivered in 1954; it seems that the 8th edition of the *Grundlagen* provided a welcome opportunity to elaborate his ideas further.

2 Freudenthal (1960, 1962, 1974).

3 Schmidt (1993, p. 406): "Anstoss zu einer neuen Art axiomatischen Denkens".

4 Here I follow the summary in Freudenthal (1960, p. 16).

5 Pasch (1882).

6 Freudenthal (1957c, p. 106).

7 Weyl (1944, p. 635), (1970, p. 164) reports that the remark probably dates from 1891, when Hilbert made it on the occasion of a lecture by Wiener on Desargues' and Pappus' theorems.

8 In quotations I shall keep to Freudenthal's German texts. In his (1962), written in English, he translated or paraphrased many passages from his (1957c) and (1960); if available I shall add these versions in a footnote; otherwise the translations in footnotes are mine.

9 Freudenthal (1960, pp. 16–17); cf. (1962, p. 619) "the turn of mathematics to axiomatics".

10 Freudenthal (1960 pp. 16–17); cf. the corresponding passage from his (1962 p. 619): "This thoroughly and profoundly elaborated piece of axiomatic workmanship was infinitely more persuasive than programmatic and philosophical speculations on space and axioms ever could be."

11 Freudenthal (1957c, p. 100).

12 Actually not all, as Freudenthal remarks in a last minute footnote (Freudenthal, 1957c, p. 112); after writing the article he found that Fano in 1892 had expressed views similar to those of Hilbert.

13 Freudenthal (1957c, p. 111); cf. (1960 pp. 617, 618): "Whereas Pasch was anxious to derive his fundamental notions from experience and to postulate no more than experience seems to grant, Hilbert started: 'Wir denken uns...' we imagine three kinds of things... called points... called lines... called planes... we imagine points, lines, and planes in some relations... called lying on, between, parallel, congruent..." (-) "Wir denken uns..." – the bond with reality is cut. Geometry has become pure mathematics. The question of whether and how to apply it to reality is the same in geometry as it is in other branches of mathematics. Axioms are not evident truths. They are not truths at all in the usual sense."

14 Freudenthal (1957c, pp. 114–115), more elaborated in Freudenthal (1960, pp. 17–25).

15 Freudenthal (1957c, p. 115); cf. (1962, p. 620): "My surprise increased further when I read all the other papers by Poincaré and the controversies they had caused. It appeared that Poincaré's reflections had almost nothing to do with the quest into the logical status of geometry, though after Hilbert's *Grundlagen* they are readily interpreted in this sense. I must confess this has been a bewildering experience for me. Historians who believe they might understand a philosopher by reading his work isolated from its environment in a more recent edition or perhaps even translated into another language should be warned. In another context a whole book may take quite another meaning – as was the case with *La Science et l'Hypothèse*."

16 Pasch (1926), original edition Pasch (1882).

17 Schmidt (1933).

18 Nagel (1939).

19 Weyl (1944).

[20] Compare his remark about the second continuity axiom: "Dies unglückselige Axiom (-) ist trotz wiederholter (auch sachlicher) Änderungen immer wieder logisch defekt formuliert worden." Freudenthal (1957c, p. 117).

[21] Freudenthal (1960, p. 5).

[22] Freudenthal (1960, p. 12).

[23] Bottema (1963 p. 168).

[24] Freudenthal (1957c, p. 2); "The totality of what these authors cooperatively fabricated".

[25] Compare Richards (1979, 1986 and 1988).

[26] Richards (1986, p. 325).

[27] In particular (Forman, 1971).

[28] Mehrtens (1990).

REFERENCES

Bottema, O.: 1963, 'Zur Geschichte der Grundlagen der Geometrie', *Mathematisch-Physikalische Semesterberichte* **9**, 164–168.

Forman, Paul: 1971, 'Weimar culture, causality, and quantum theory, 1918–1927: adaption by German physicists and mathematicians to a hostile intellectual environment', *Historical studies in the physical sciences* **3**, 1–115.

Freudenthal, Hans: 1957a, *Neuere Fassungen des Riemann- Lieschen Raumproblems*, in *Der Begriff des Raumes in der Geometrie* Naas, Schröder, (eds.), Berlin, (pp. 92–97).

Freudenthal, Hans: 1957b, 'La topologie dans les fondements de la géometrie.' in J.C.H. Gerretsen and J. de Groot (eds.), *Proceedings of the international congress of mathematicians 1954* Groningen etc., Noordhoff etc., vol. 3, pp. 178–184.

Freudenthal, Hans: 1957c 'Zur Geschichte der Grundlagen der Geometrie – zugleich eine Besprechung der 8. Aufl. von Hilberts Grundlagen der geometrie', *Nieuw Archief voor Wiskunde*, (3) 5 105–142.

Freudenthal, Hans: 1960 'Die Grundlagen der Geometrie um die Wende des 19. Jahrhunderts', *Mathematisch-Physikalische Semesterberichte* N.F. 7, 2–25.

Freudenthal, Hans: 1962, 'The main trends in the foundations of geometry in the 19th century', in E. Nagel e.a. (eds.), *Logic Methodology and Philosophy of Science, Proceedings of 1960 International Congress*, Stanford, 613–621.

Freudenthal, Hans: 1974, 'The impact of Von Staudt's foundations of geometry', in R.S. Cohen, J.J. Stachel (eds.), *For Dirk Struik*, Reidel Publ. Co., Dordrecht, pp. 189–200.

Hilbert, David: 1956, *Grundlagen der Geometrie* (Rev. add. P. Bernays), 8th ed., Teubner, Stuttgart.

Mehrtens, Herbert: 1990, *Moderne – Sprache – Mathematik; Eine Geschichte des Streits um die Grundlagen der Disziplin und des Subjekts formaler Systeme*, Suhrkamp, Frankfurt.

Nagel, E: 1939, 'The formation of modern conceptions of formal logic in the development of geometry', *Osiris* **7**, 142–224.

Pasch, Moritz: 1882, *Vorlesungen über neuere Geometrie*, Leipzig.

Pasch, Moritz and Max Dehn: 1926, *Vorlesungen über neuere Geometrie* (with appendix *Die Grundlagen der Geometrie in historischer Entwicklung*, by Max Dehn), Berlin.

Richards, Joan L.: 1979, 'The reception of a mathematical theory: non-Euclidean geometry in England 1868–1883' in B. Barnes and S. Shapin (eds.), *Natural order; historical studies of scientific culture*, Sage, London, pp. 143–166.

Richards, Joan L.: 1986, 'Projective geometry and mathematical progress in mid-Victorian Britain', *Studies in history and philosophy of science* **17**, 297–326.

Richards, Joan L.: 1988, *Mathematical visions – the pursuit of geometry in Victorian England*, Academic Press, New York.

Schmidt, Arnold: 1993, 'Zu Hilberts Grundlagung der Geometrie' in David Hilbert, *Gesammelte Abhandlungen* (3 vols., Berlin 1932–1935, reprint New York 1965), 2, 404–414.

Weyl, Hermann: 1944, 'David Hilbert and his mathematical work', *Bulletin of the American Mathematical Society* **50**, 612–654.
Weyl, Hermann: 1970, 'David Hilbert and his mathematical work', in Constance Reid, *Hilbert*, Springer, Berlin, pp. 245–283.

Mathematical Institute, Utrecht University
Budapestlaan 6
3584 CD Utrecht, The Netherlands

Never before have I seen so much of the world and so many people as from 9 July to the 15th (?) of August 1921 (diary 1184–1289). It started in Freiburg i.B., where one of our former borders, Max Lefkowitz, was continuing his medical studies begun in Berlin. I attended several classes – I thought the best to be the medical ones. The scientific I described as of low level. I noticed that Husserl gesticulated a lot, that he repeated everything ten times as if he were speaking to the mentally retarded and that he distorted the entire history of philosophy phenomenologically. On 15 July I wrote down: "Again he talked about the good and reputable Locke, who is good and reputable enough to ... he ended the class with a sentence in which the word 'Widersinn'* occurred ten times".

from: H.F.: Schrijf dat op Hans. Knipsels uit een leven. Meulenhoff, Amsterdam, 1987, pp. 110–11.

* Means: nonsense.

W. T. VAN EST

HANS FREUDENTHAL*

(17 September 1905–13 October 1990)

ABSTRACT. Recorded are some biographical data about the late professor Freudenthal along with some indications about his mathematical work. In an appendix a tiny part of his mathematical work which can be explained in a fairly direct manner is discussed.

There are scientists and scholars. Hans Freudenthal was a scholar. There are those whose activities are confined within the limits of their discipline and there are the ones whose activities extend far beyond these boundaries. Hans Freudenthal belonged to the latter category. Like few others he proved capable of synthesizing an extensive range of knowledge with mathematics as the dominant key. At the same time this went hand in hand with activities on a broad scale, essays, literary work, innovation of mathematics education, and numerous major and minor activities for which an appeal was made to him, seldom in vain. That all this did not go unnoticed by the Dutch mathematical community, has been proven repeatedly on the occasions of his sixtieth, seventieth, and eightieth birthday, as well as by the honorary membership of the Wiskundig Genootschap and the honorary doctorate of the University of Amsterdam. The surveys of his work presented on those occasions and the articles on his work in the Nieuw Archief voor Wiskunde 8 (1990), 9–12, and 9 (1991), 129–212, permit me to restrict myself to recording a few of the earlier, and therefore perhaps less known episodes, in the life of Freudenthal.

Hans Freudenthal was born in Luckenwalde on 17 September 1905, the son of Elisabeth Ehmann and the teacher Joseph Freudenthal. He attended the local Reformrealgymnasium and registered as a student at the university of Berlin in 1923.

The relatively short distance between Luckenwalde and Berlin permitted him initially to commute by train until he found a room in Berlin from where he would go home for the weekends, returning on Sunday with his laundry done – a student life style not so uncommun even today.

Freudenthal always remembered his student days at the University of Berlin as an unforgettable experience[1]. The atmosphere of the mathematics institute and the cordial relationship among students and between students and professors remained ideal in his mind. In his later years, as a professor, Dutch reality did not always

Educational Studies in Mathematics **25**: 59–69, 1993.

quite correspond to this ideal, but in any case he experienced the satisfaction of seeing his ideal more or less approached.

Besides Göttingen, Berlin was another centre of importance with professors such as Erhard Schmidt, von Mises, Isaî Schur, Bieberbach, and with Loewner, von Neumann and Heinz Hopf among the younger lecturers. In physics Planck and Einstein attracted full audiences. Visitors form abroad were no rarity in Berlin. A visit by Brouwer in the winter semester '26–'27 brought Freudenthal into closer contact with intuitionism – a first, not entirely successful introduction he had had in an analysis course by Loewner. And it was through intuitionism that he was recruited in 1930 by Brouwer as an assistant.

After having spent the summer semester of 1927 in Paris where he heard Hadamard, Julia and Picard, he returned to Berlin as a "Hilfsassistent am Mathematischen Seminar", thus affording him an opportunity to prepare a Ph.D. As Freudenthal once wrote later, of course one wanted to do work on one of the "hot topics" of the day – another life style phenomenon that still persists.

Since at that time topology was beginning to bear its first fruits, it was preeminently an area that offered scope to this ambition. The result was a thesis about Ends of Topological Spaces and Topological Groups[2], with Hopf as thesis adviser and Bieberbach formally as thesis director, which was presented at the Doktorexamen in 1930. The formal granting of the degree did not take place until 1931 when the thesis was accepted as an article by a mathematical journal.

It seems appropriate to say a few words about the notion of "Freudenthal-end" that has since gained a permanent position in topology.

As explained in textbooks on topology, a locally compact space admits a compactification by adding one point at infinity, the so-called Alexandroff-point. Open neighbourhoods of this point are by definition the union of the Alexandroff-point and the complement of a compact set in the original space. Thus the real line is compactified into a circle, and an infinite cylinder is turned into a torus with one meridian shrunk to a point.

Suppose that for a fixed $k \geq 1$, there exist arbitrarily small neighbourhoods of the Alexandroff-point that can be split up into k relatively closed connected pieces $C_1, \ldots, C_k$ that have pairwise only the Alexandroff-point in common and each of which intersects the complement of the Alexandroff-point in an open connected set. Select one such neighbourhood, then by judiciously cutting up the Alexandroff-point, so that the pieces $C_1, \ldots, C_k$ come apart from one another, one obtains a new compact space. The original Alexandroff-point is cut up into k copies, each of the $C_1, \ldots, C_k$ containing exactly one such copy. These k copies of the original Alexandroff-point are the ends in the sense of Freudenthal. Thus the real line is compactified into a closed interval, and the infinite cylinder into a 2-sphere: in both cases there are exactly two Freudenthal-ends. In general: parts of the original space which were separated by a compact set remain separated by the same compact set in the Freudenthal-compactification.

A substantial part of Freudenthal's thesis is devoted to putting this simple intuitive idea on a firm basis. The second main result is that a locally compact group has 0, 1 or 2 ends; zero ends only in the event of compactness. For further developments of the theory see notes 3–5.

As mentioned before it was his interest in intuitionism rather than in topology that was the main reason for Freudenthal's appointments as an assistant to Brouwer in 1930. In Amsterdam Freudenthal met Hurewicz, another assistant of Brouwer, who had come to Amsterdam a few years earlier precisely through his interest in topology. Although Brouwer was no longer actively interested in topology at that time, both, as Freudenthal would later write, considered their assistantship as an honour and a challenge. This soon became apparent by the special courses they organized with great enthusiasm, formally entitled to do so by being appointed "privaat-docent" – a position without remuneration that only carried the right to lecture. Among the courses they gave we find: Algebraic Fields (elementary theory), Group Theory, Measure Theory, Complex Analysis, Topology, Linear Operators, Elementary Mathematics from an Advanced Standpoint; in short, optional topics that stood apart from the standard mathematics curriculum at the time.

Also scientifically the Amsterdam period was a fruitful one for both of them. Hurewicz developed his theory of Homotopy Groups, a new tremendous step in topology after the work of Poincaré and Brouwer. Immediately Freudenthal made an important contribution to this theory by proving the now classic suspension theorems. Furthermore Freudenthal published papers on Topological Groups, Lie Groups, Almost Periodic Functions, Algebraic Topology, Functional Analysis, and Intuitionism, and these soon established his reputation within the international mathematical community.

The year 1937 brought some major changes. The professors Mannoury and de Vries (Hk) retired, Hurewicz had left for the U.S. in 1936. Heyting and Freudenthal were to take over the regular teaching duties of the retired professors and were appointed as lecturer and curator respectively; the crisis of the thirties still called for budget cuts.

Heyting who was charged with teaching Geometry and Algebra, gave a second year course on modern algebra, based on the book by Van der Waerden, concluding with Galois Theory. For the students this course opened up an entirely new world. For the third and fourth year students he gave a modernized course on differential geometry inspired by the booklet on differential geometry by Bieberbach.

On the other hand the analysis courses by Freudenthal underwent a drastic change in method. The first and second year courses could be characterized as being in the spirit of Bourbaki avant la lettre. The number concept was developed from the Peano axioms for natural numbers up to the real and complex numbers. The complex domain was not shunned whenever it appeared appropriate to avail oneself of it, e.g. in the treatment of power series, and in d'Alembert's proof of the main theorem of algebra. The language of sets was used from the outset, and metric spaces were introduced as a natural setting for some basic theorems in analysis. Analysis of several variables was developed within the framework of normed (finite dimensional) linear spaces. Differential forms were introduced and the generalized Stokes theorem was proved. The theory of Lebesgue integration concluded this two year course.

This abstract approach initially did cause some difficulties in understanding on the part of the students. However, on the whole one must say that students were quite impressed by the course irrespective as to whether they had mathematics was their first or second major.

The follow up in the later years consisted of the Theory of a Complex Variable, Differential Equations (ordinary and partial), Functional Analysis (Banach spaces, spectral theorems, Stone–von Neumann for one parameter unitary groups, ergodic theory, almost periodic functions).

Of course both Heyting and Freudenthal lectured in addition on topics that had their special interest, such as Foundations of Mathematics and Intuitionism (Heyting), and Topology (Freudenthal).

The war brought an abrupt end to this relatively care-free period. Like many others, Freudenthal was suspended from duties. This situation allowed on the one hand plenty of time for personal work – historical studies and some work published after the war were fruits of this period. Furthermore there was the satisfaction of seeing two young mathematicians, J. de Groot and A. van Heemert taking degrees with Schaake at Groningen; de Groot's thesis[6] treated a topic closely related to the theory of ends, whereas van Heemert's thesis[7] applied in an essential way the theory of projective and inductive limits developed by Freudenthal[8].

On the other hand, however, the situation was too threatening and even more so as the war progressed, to permit real "enjoyment" of this forced "leave of absence". It is probably due for a considerable part to the efforts of Mrs Freudenthal, who where possible acted in defence of her husband, that the Freudenthal family was spared from the gravest ordeal. A few arrests of short duration, a period in a work camp in Holland, and a period of hiding out in the south of the country, may after all be considered as a happy ending. The help that Freudenthal received from students in Amsterdam should not go unmentioned.

After the war the faculty of sciences at Utrecht approached Freudenthal, offering him the chair for Geometry (fallen vacant by the retirement of professor Barrau). There was no immediate prospect for a professorship at the University of Amsterdam. An appealing aspect of the Utrecht offer was the possibility to modernize the mathematics curriculum in collaboration with Popken, the candidate for the Analysis chair[10]. This made the transition easier and a new period began.

The teaching assignment of Geometry aroused the interest in the connection between geometries and their symmetry groups, a theme that has constantly been recurring in mathematics ever since the formulation of the so-called Erlanger Programm in 1872 by Felix Klein[9] in which precisely the connection between geometries and their symmetry group was suggested as a theme for future research.

The interest in this theme at Utrecht broadened gradually and a "Utrecht School" of Geometry and Lie Groups developed.

Active participation since 1948 (and even earlier) in the innovation of mathematics education in the "Werkgemeenschap voor Vernieuwing van Opvoeding en Onderwijs" demanded increasingly more time in the later years. As a result of this the IOWO was founded, the institute for the development of (new methods in) mathematics education, later to become part of the active Research Group on Mathematics Education and Educational Computer Science at the University of Utrecht.

On the occasion of his eightieth birthday the exceptional contribution of Freudenthal to the innovation of mathematics education was extensively commemorated.

The merits of Freudenthal for the Wiskundig Genootschap and the Dutch mathematical community at large were many. He served the Genootschap twice as a president, and played an important role in bringing about a number of major changes in the life of the Society.

Just to mention one of his other merits one might recall his role (with others) in reactivating in the fifties the mathematical journal Compositio Mathematica.

Freudenthal was active up to his very last day. He passed away peacefully, sitting on a park bench on a sunny October morning. The Wiskundig Genootschap commemorates him as one of the outstanding members in its history.

APPENDIX

This appendix is meant to explain a few of the topics on which professor Freudenthal worked. We have chosen Homotopy Theory and Lie Groups in Geometry as subjects, because these are the ones with a suggestive geometric content and therefore may be explained in a relatively non-technical manner.

Homotopy Theory

One of the oldest notions in topology is that of *connectedness*. We say that a space X is *connected* – and the notion of (topological) space is taken in the sense of a "space" for which the notion of a continuous map from a subset of euclidean space into the "space" in question makes sense – if any pair of points p, q in X can be joined by a continuous curve, or *path*, i.e. if there exists a continuous map $f\colon [0, 1] \to X$, such that $f(0) = p$, and $f(1) = q$, the *initial point* and the *end point* respectively.

Strictly speaking in topology this is called pathwise connectedness of the space; we shall, however, stick to the term connected(ness).

Another well known notion is that of *simple connectedness*. Recall for instance that in the theory of a complex variable, in order to show that an analytic function ϕ defined in a certain domain X has a primitive, one fixes a point o in X and one defines the value in a point q of the sought for primitive as the integral $\int \phi(z)\,\mathrm{d}z$ along a path joining o with q. As soon as for any two paths f_0, f_1 there exists a continuous deformation sweeping f_0 onto f_1, keeping the extremal points fixed, then, by the analyticity of ϕ, it follows that the two integrals $\int \phi(z)\,\mathrm{d}z$ yield the same value, and hence the primitive is well defined.

Now in general a connected space X will be called simply connected if for any pair of points p, q in X and any pair of paths f_0, f_1, joining p and q there exists a *deformation*, i.e. a parametrized family of paths f_s, $0 \leq s \leq 1$, such that $f_s(t)$ is continuous in the pair (s, t) and that $f_s(0) = p$, $f_s(1) = q$ for all s, and yielding for $s = 0$ the path f_0 and for $s = 1$ the path f_1.

In particular if X is simply connected, then choosing a base point p in X, any *loop* in X based at p, i.e. any path with p both as initial and end point, admits a deformation within the space $\Omega(X, p)$ of loops based at p into the constant loop $\tilde{p}$ with $\tilde{p}(t) = p$, $0 \leq t \leq 1$. It is easily shown that this property for a single point p in X already implies simple connectedness for X.

Now a continuous deformation f_s of loops based at p may be seen as a path in $\Omega(X, p)$. Hence simple connectedness of X just means that the "space" $\Omega(X, p)$ is connected. In a non-simply connected space X the loop space $\Omega(X, p)$ decomposes into various connectedness components.

Well known examples of simply connected spaces are convex subsets of euclidean spaces. Furthermore a sphere of dimension ≥ 2 is shown to be simply connected. However, the 1-sphere, i.e. the circle, and likewise the euclidean plane minus the origin are examples of non-simply connected spaces.

In the last example it is almost intuitive that the connectedness components of the space $\Omega(X, p)$ are completely classified by their winding number, i.e. the number of times the loop encircles the left out origin, this number being counted positive if in total the loop encircles counter clockwise and zero or negative otherwise. Hence we may say that the connectedness components are classified by the integers. There is, however, more to be said.

If we compose two loops f and g into one loop by first describing the loop f and subsequently g, "speeding up at the same time the velocity" so that the combined loop $f \vee g$ again takes only one second in time, we see that the winding numbers of f and g add up to the winding number of $f \vee g$. In other words, the composition of loops just described induces a composition on the set of connectedness components so as to turn it into a group isomorphic to the additive group of integers.

Now this is a general phenomenon. The composition of loops induces on the collection of connectedness components of a loop space $\Omega(X, p)$ a group law; the resulting group is called the *first homotopy group* $\pi_1(X, p)$ of X based at p, or *fundamental group*. As a matter of fact there is a slight difference between the notions of first homotopy group and fundamental group, but usually in literature they are hardly distinguished (if at all) and we shall do likewise.

The notion of fundamental group was introduced by Poincaré in 1895 and it plays a role among others in the theory of algebraic curves, in the theory of Lie groups and various other parts of mathematics.

Forty years later Hurewicz came up with the enlightened idea to study the fundamental group $\pi_1(\Omega_0(X, p), \tilde{p})$ of the connectedness component $\Omega_0(X, p)$ of the constant loop in the space $\Omega(X, p)$; this group was called the *second homotopy group* $\pi_2(X, p)$ of X based at p. But then of course one can continue in this fashion defining $\pi_1(\Omega_0(\Omega_0(X, p)), \tilde{\tilde{p}})$ to be the third homotopy group $\pi_3(X, p)$ of X based at p, etc. Of course for these definitions to make sense one has to say how paths and deformation of paths in these spaces are to be defined, but these definitions are completely patterned after the fashion we defined paths in $\Omega(X, p)$.

At the same time it was shown that the groups $\pi_n(X, p)$ could also be defined directly starting from "spherical n-loops". To that end mark a base point o on the n-sphere S^n, i.e. the boundary of the unit ball in euclidean $(n + 1)$-space. Now consider the set $\Sigma^n(X, p)$ of continuous maps $S^n \to X$ carrying o onto p. A deformation of such maps is defined just as for loops, and thus a deformation may again be considered as a path in $\Sigma^n(X, p)$. Hence in general $\Sigma^n(X, p)$ will decompose into various connectedness components. Furthermore in order to define a composition $f \vee g$ for two such maps $(S^n, o) \to (X, p)$ we proceed as follows: First we squeeze a great $(n - 1)$-sphere through o to a single point, so that S^n is transformed into a couple of n spheres attached to each other in just one point that we continue to denote by o. Then map one member of this couple by f and the other one by g; this will then yield a map of the original (S^n, o) into (X, p).

This composition of spherical n-loops again induces a composition on the set of connectedness components turning this set into a group, which, as may be guessed from the construction, is a commutative group for $n \geq 2$. Now these groups identify in a natural fashion with the homotopy groups $\pi_n(X, p)$.

The definition of homotopy groups by means of spherical loops had been presented independently by E. Cech at the Zürich congress of mathematicians in 1932. However, Cech did not work out his ideas.

In any case Hurewicz proceeded to relate the homotopy groups to the so called homology groups introduced by Poincaré. One striking result obtained by Hurewicz was that in a space where the homotopy groups of order ≥ 2 vanish, the homology groups are entirely determined by the fundamental group. This result was one of the germs of the branch of mathematics called homological algebra that developed much later.

The relationship between homotopy and homology developed by Hurewicz confirmed that for the n-sphere S^n, $\pi_i(S^n, p) = 0$ for $i < n$, and $= \mathbb{Z}$ for $i = n$, a result that was obtained earlier in a somewhat different formulation. Another striking fact of this theory was that it gave a very elegant and simple proof of the fact that $\pi_3(S^n, p) = \mathbb{Z}$, again a result that was essentially known, but with a proof that was more involved.

This centred the attention to the homotopy groups of spheres, and here Freudenthal obtained a remarkable result, to wit $\pi_i(S^n) = \pi_{i+1}(S^{n+1})$ (we omit base points) for $i \leq 2n - 2$, or writing $i = n + d$, $\pi_{n+d}(S^n) = \pi_{n+1+d}(S^{n+1})$ for $d \leq n-2$. Obviously this result expresses a certain stability property of homotopy groups: the homotopy group of a sphere in a dimension that exceeds the sphere dimension by a fixed d remains constant, once the condition $d \leq n-2$ is satisfied.

So for example Freudenthal proved that $\pi_4(S^3) = \mathbb{Z}_2$ – the 2-cyclic group – and hence it follows that $\pi_{n+1}(S^n) = \mathbb{Z}_2$ for $n \geq 3$.

Much later many more stability results of a similar type were discovered. In any case this result stood apart as a remarkable one for a long time, till in the fifties the subject began to move again, and one might say that by now the Freudenthal result is well understood from various points of view.

Lie Groups

The prototype of a Lie group is the group of non-singular $n \times n$-matrices, to be denoted by GL(n, $\mathbb{R}$) in case we are dealing with real matrices, and by GL(n, $\mathbb{C}$) if the matrices are supposed to be complex. However, we shall often write GL(n) leaving undetermined whether we are dealing with $\mathbb{R}$ or $\mathbb{C}$.

Now GL(n) is an open subset of the n^2-dimensional space of all $n \times n$-matrices; the matrix elements serve as standard parameters. Multiplication and inversion are expressed by rational functions in terms of the argument parameters involved and therefore depend analytically on the arguments.

In general a *Lie group* is a manifold on which a group law is defined in such a manner that multiplication and inversion are expressed by analytic functions in terms of the parameters involved. The Lie group is said to be *real* if the manifold on which the group law is defined is a real manifold (the local parameters being real numbers) and *complex* if the manifold is a complex one (the local parameters being complex). Subgroups of GL(n) that are singled out by polynomial equations in the parameters are shown to be Lie groups. So for instance the group of SL(n) of $n \times n$-matrices of determinant 1 is a Lie group, and likewise SO(n), the group of orthogonal matrices of determinant 1, and Sp(n) the group of $2n \times 2n$-matrices that leave invariant the skew symmetric bilinear form $\Sigma_{i=1}^{n}(x_i y_{n+i} - x_{n+i} y_i)$, are Lie groups. Again we write SL(n, $\mathbb{R}$) or SL(n, $\mathbb{C}$), and similarly for the other groups, according to the case that we are dealing with real or complex matrices. If not mentioned otherwise, homomorphisms, isomorphisms, automorphisms of Lie groups are understood to be analytic.

Apart from the notion of isomorphism there is also the notion of *local isomorphism* in Lie group theory. We say that the Lie groups G and H are locally isomorphic if there exists an analytic bijection ϕ from a neighbourhood U of the identity in G onto a neighbourhood V of the identity in H such that $\phi(u_1)\phi(u_2)$ is defined in V whenever $u_1 u_2$ is defined in U and vice versa, and furthermore $\phi(u_1 u_2) = \phi(u_1)\phi(u_2)$. The notion of local isomorphism is an equivalence notion. One proves that, in the category of *connected* Lie groups, in an equivalence class of locally isomorphic Lie groups there is up to isomorphism just one simply connected Lie group such that every other member of the class is (up to isomorphism) the quotient of this simply connected one by a discrete central subgroup.

From now on we restrict our attention to the category of *connected* Lie groups.

In analogy with the case of finite groups a Lie group is said to be *simple* if the only (connected!) normal subgroups are the identity and the total group. If a Lie group is simple, so is any group locally isomorphic with the given one.

The classification of complex simple Lie groups was obtained long ago by Lie and Killing, and up to local isomorphism they are the groups SL(n, $\mathbb{C}$), SO(n, $\mathbb{C}$),

Sp(n, $\mathbb{C}$) and furthermore five exceptional types that are denoted by G_2, F_4, E_6, E_7, E_8.

In the literature the series SO(n, $\mathbb{C}$) is usually split up into two series according to n even or odd, so that one speaks of the four main series and the five exceptional Lie groups.

Then as far as the real Lie groups are concerned there is a similar classification. First of all considering the complex simple groups as real groups of real dimension twice the complex dimension of the group, the complex simple Lie groups are at the same time simple real Lie groups. But apart from these, there still are the real Lie groups which are the real counterparts of the complex simple Lie groups, and whose real dimension is equal to the complex dimension of the corresponding complex simple group; these groups are the so-called *real forms* of the complex ones. So e.g. SL(n, $\mathbb{R}$) is a real form of SL(n, $\mathbb{C}$), but also the special unitary group SU(n), the group $n \times n$-unitary matrices of determinant 1, is a real form of SL(n, $\mathbb{C}$). The real forms of SO(n, $\mathbb{C}$) are the groups SO(p, q), $p + q = n$, of the real $n \times n$-matrices that leave invariant a quadratic form $x_1^2 + x_2^2 + \cdots + x_p^2 - x_{p+1}^2 - \cdots - x_n^2$.

One of the problems Freudenthal was concerned with was the relationship between the topology of a Lie group and its abstract group structure. This question is intimately related to the following one: Given an abstract group isomorphism of Lie groups, under what conditions will it be analytic?

Observe for instance that any field automorphism $\phi : \mathbb{C} \to \mathbb{C}$ of the complex numbers will yield an automorphism of SL(n, $\mathbb{C}$) by applying ϕ to the matrix elements. Since there exist highly discontinuous automorphisms of the field of complex numbers, there exist discontinuous abstract group automorphisms of SL(n, $\mathbb{C}$); a similar observation holds for SO(n, $\mathbb{C}$) and Sp(n, $\mathbb{C}$).

Now Freudenthal proved that any isomorphism of a *real form* of a complex simple Lie group is necessarily analytic. This result extended an earlier result of E. Cartan and B. L. van der Waerden.

Again this result stood apart for a long time, till in the seventies the question was taken up again and reformulated in the framework of algebraic linear groups by Borel and Tits.

Another theme in the theory of Lie groups that attracted Freudenthal's attention was that of the relationship between geometries and their symmetry groups.

Recall that by the second half of the 19th century a great deal was known about geometries such as euclidean geometry, affine geometry, projective geometry, non-euclidean geometries, line geometry, circle geometry, etc. Felix Klein in his so-called Erlanger Programm observed that all these geometries were in a certain sense characterized by their symmetry groups in that the mutual relationships between the objects of these geometries and their properties were invariant under the symmetry group of the geometry, which always happened to be a Lie group. And he formulated as a project of study the following: Given a manifold with a transitive Lie group of transformations. Develop the theory of mutual relationships between objects of this manifold and their properties that are invariant under the given group of transformations.

One could go even one step further in formulating the following project: Given a Lie group a priori. Find the geometry(ies) for which this group is the symmetry group.

Now Freudenthal's interest in this field awoke when he was lecturing (for many years) on projective geometry. Recall that for any field, and even any skew field, one constructs a projective plane over that field by a standard procedure. So one has the real projective plane, the complex projective plane, and the projective plane over the quaternions. In any of these planes the objects to be studied are points, "straight lines", their incidence relations and the properties of the figures made up of points and lines. Of course straight lines seen as sets of points look topologically markedly different from one type of geometry to another. In the real projective plane they are topological circles, in the complex projective plane they are topological 2-spheres, and in the geometry over the quaternions they are 4-spheres. Apart from the classical fields of the reals, the complex numbers, and the quaternions, there is the algebra of the so-called octonions or Caley numbers which is not a field since the multiplication is not associative, but in which every non-zero element has an inverse (like in the case of a field).

Earlier results in the literature made it plausible that also over the Caley numbers a projective plane could be constructed albeit by a procedure different from the standard one. Let us just state for clarity that the notion of projective plane is that of a geometry in which the basic elements are *points* and *lines* and a relationship of *incidence*. The basic axioms relating these notions are then:

1. There is exactly one line incident with two given points.
2. There is exactly one point incident with two given lines.

In any case Freudenthal gave a construction of a projective plane over the Cayley numbers, and established a number of properties of this geometry. The symmetry group turned out to be E_6. Furthermore, just as in the case of the real projective plane that serves as a model for the elliptic non-euclidean geometry by introducing a suitable metric, the Cayley plane admits a similar geometry, and then the symmetry group reduces to F_4.

Then finding suitable geometries for E_7, E_8 was still a problem. This was solved to a great extent by Freudenthal and his students.

By now it is clear that the Klein-project has found a solution in great generality by the work of J. Tits in his theory of Buildings. We will not enter into that, however.

NOTES

* This is the somewhat modified text of a commemorative address given at the annual meeting of the Wiskundig Genootschap, the Dutch mathematical society, on April 5, 1991, cf. Nieuw Archief voor Wiskunde, 9 (1991), 131–136.

[1] Freudenthal, H.: 1987, *'Berlin 1923–1930, Studienerrinerungen von Hans Freudenthal'*, DMV-Tagung 21–25 September 1987, Sonderdruck, W. de Gruyter, Berlin.

[2] Freudenthal, H.: 1931, 'Über die Enden topologischer Räume und Gruppen', *Mathematische Zeitschrift* **33**, 692–713.

[3] Peschke, G.: 1990, 'The Theory of Ends', *Nieuw Archief voor Wiskunde* **8**, 9–12.

[4] Specker, E.: 1950, 'Endenverbände von Räumen und Gruppen', *Mathematische Annalen* **122**, 167–174.

[5] Borel, A.: 1953, 'Les bouts des espaces homogènes de groupes de Lie', *Annals of Mathematics* **58**(2), 443–457.

[6] Groot, J. de: 1942, *Topologische Studiën*, Ph.D. thesis R.U. Groningen.

[7] Heemert, A. van: 1943, *De R_n-adische Voortbrenging van Algemeen Topologische Ruimten met Toepassingen op de Constructie van niet-splitsbare Continua*, Ph.D. thesis R.U. Groningen.

[8] Freudenthal, H.: 1937, 'Entwicklungen von Räumen und ihren Gruppen', *Compositio Mathematica* **4**, 145–234.

[9] Klein, F.: 1921, 'Vergleichende Betrachtungen über neuere geometrische Forschungen', *Ges. Math. Abhn.* **I**, 460–497.

[10] This chair had fallen vacant by the tragic death of professor J. Wolff in the concentration camp Theresiënstadt. Likewise Dr. L. W. Nieland, who most generously had taken over the teaching duties of professor Wolff for the time being, refusing any remuneration for it, fell victim to another war drama.

REFERENCES

Freudenthal, H.: 1937, 'Über die Klassen der Sphärenabbildungen I', *Compositio Mathematica* **5**, 299–314.

Freudenthal, H.: 1941, 'Die Topologie der Lieschen Gruppen als algebraïsches Phänomen, I', *Annals of Mathematics* **42**, 1051–1074.

Freudenthal, H.: 1985, 'Oktaven, Ausnahmegruppen und Oktavengeometrie', *Geometriae Dedicata* **19**, 1–63 (reprint of a mimeographed report of 1951).

Hurewicz, W.: 1935, 'Beiträge zur Topologie der Deformationen, I, II', *Proceedings of the KNAW* **38**, 112–119, 521–528.

Hurewicz, W.: 1936, 'Beiträge zur Topologie der Deformationen, III, IV', *Proceedings of the KNAW* **39**, 117–126, 215–224.

Cech, E.: 1932, 'Höhere Homotopiegruppen', *Verh. Int. Math. Congr. Zürich* **II**, 203.

Cartan, E.: 1930, 'Sur les représentations linéaires des groupes clos', *Commentarii Mathematici Helvetici* **2**, 269–283.

Borel, A. et Tits, J.: 1973, 'Homomorphismes abstraits de groupes algébriques simples', *Annals of Mathematics* **74**, 499–571.

Klein, F.: 1921, 'Vergleichende Betrachtungen über neuere geometrische Forschungen', *Ges. Math. Abhn.* **I**, 460–497.

Tits, J.: 1974, *Buildings of Spherical Type and Finite BN-Pairs*, Springer, Berlin etc. Lecture Notes in Mathematics 386.

Van der Waerden, B. L.: 1933, 'Stetigkeitssätze für halbeinfache Liesche Gruppen', *Mathematische Zeitschrift* **36**, 780–796.

Aert van Neslaan 628
2341 HV Oegstgeest,
The Netherlands

HANS FREUDENTHAL

THOUGHTS ON TEACHING MECHANICS
DIDACTICAL PHENOMENOLOGY OF THE CONCEPT OF FORCE

If it is true – and who would deny it? – that in mechanics instruction body experiences interfere inconveniently with scientific ideas, then it is of paramount importance, instead of suppressing them, to have the learning processes started just there, and the learner, under guidance, transform them into what we consider scientific.

INTRODUCTION

When dealing with education I seldom ventured outside mathematics, but if I did the subject area was very likely to be mechanics[1]. The present paper is more principles-orientated than the previous ones. The incubation period has been considerable; eventually its first edition[2] has been incited by an Utrecht Ph.D. thesis[3], connected to the remarkable developmental work of the Dutch group PLON[4].

I cannot judge to which degree textbooks I am familiar with are representative of science teaching in general, and the same holds true for research. Almost all of it concerns *errors* (or are these errors?), committed by young and old people who did not, or even who did, study some science. If not created, this trend was reinforced by Piaget when he renounced his initial marvellous positive inquiry into cognitive learning in favour of a rather negative approach.

Rarely, if ever, do researchers ask, let alone investigate, whether and to which degree "errors" are due to education, and what educational developers and teachers can learn from them, although only tackling these questions could justify this kind of research as properly educational.

No doubt, readers at home in the field in question, will immediately remember the hundreds of – virtually identical – investigations around the concept of force: Illustrated by a drawing which shows a cyclist on his vehicle and opposite arrows on the ground which are to represent forces, such test questions are posed as to which force is greater, the forward or backward one (of friction); or for a change, similar drawings and questions about a pendulum being swung, or the

Educational Studies in Mathematics **25:** 71–87, 1993.

parabolic trajectory of a ball tossed up. The hypothesis that tests will give "wrong" answers is amply confirmed, from kindergarten to college and even faculty. To be sure, hunting for errors has, most of the time, been a favourite sport in research on mathematics education too, but its reputation, though still high, has been declining for a few years. The idea that mathematics, as a teaching/learning subject, differs from all the other ones fundamentally, is as old as the socratic lesson in Plato's "Menon", if not older. At various opportunities[5] I have pointed out that, invented and reinvented on a manifold of places on earth independently, mathematics, unlike any other science, has been and still is a matter of *common sense*: in the course of individual histories and that of mankind, gradually refined common sense. So didactically, it seems to be no exaggerated requirement to have this knowledge *reinvented* by the learner, albeit under *guidance*. All the same, Archimedes' principle, the law of inertia, the theory of gravitation, Proust's law have been invented once and only once. Hence this kind of knowledge seems less accessible to reinvention in learning processes. Physics and chemistry are no common sense; they are rather contrary to common sense. It is the teacher's task to fight common sense by exorcising it: common sense concepts obstruct scientific ones. I have often heard this postulate – the strongest at a 1976 conference in Rome[6]. I was not convinced then, and every time I thought it over, my doubts increased.

Modern textbooks seem to prefer physics taught within contexts. Instruction within a context has also been developed for mathematics, but in the designs I am most familiar with, the function assigned to the context differs basically from what it seems to be in teaching physics. A key concept in nowadays views on mathematical invention is *mathematising*, which means, turning a non-mathematical matter into mathematics, or a mathematically underdeveloped matter into more distinct mathematics. In mathematics instruction as I have in view, the context is, in the first instance, justified as an opportunity for the learner to mathematise it: mathematising as a learning matter is an aim in itself (which includes higher level mathematisations of contexts previously enriched with mathematics). "In the first instance", I said, since on the rebound, contexts become domains of application.

The concept of mathematising has been refined by Treffers[7], who has distinguished two components: *horizontal*, that is, from the world in which the learner lives to what is still a world of symbols to him; and *vertical*, that is, within the world of symbols; moreover, according to the presence or absence of one of these components, Treffers has distinguished four "ideal types" of education.

Does science in fact contradict common sense? Has much of science, in particular mechanics, not already become common sense? (Think, for instance, of some astronomical insights, such as shape and motion of the Earth!) Indeed, among the sciences, mechanics is the closest to mathematics, in particular to geometry. Should we not, as a counterpart to mathematising, admit "*mechanising*" or (to be applied it in a more general way) subject-area structuring? It is a needless question, since mechanising exists anyway, at least as an activity of first inventors and researchers. The real question aims at a reinventing activity allowed to the learner.

If, in teaching mechanics, the context would be attributed the function to be mechanised by the learner, it would be as well reasonable to consider mechanising refined à la Treffers: *horizontal*, from the world of life to the world of symbols; and *vertical*, within the world of symbols. As far as I am familiar with nowadays instruction in mechanics, I can neatly distinguish two types: *empiristic*, with rich opportunities for horizontal mechanising and none for the vertical component; and *structuralist*, with the opposite features. There is, to my view, all reason to create a *realistic* counterpart, which would account equally for both components of mechanising.

I have, with a view on mathematics, repeatedly pleaded for *didactical phenomenology*; eventually I have tried myself to supply the need[8]. Would something similar not be useful in other fields? Mechanics can boast of examples of *historical* and *epistemological* (for instance, Mach 1912) but, as far as I know, none of *didactical* phenomenology. In the literature I have consulted I could not even trace such implicit influences of phenomenology as one might readily expect in the case of "*force*" where they are urgently needed. As a consequence, textbooks of the structuralist type are burdened with access on, to my view, much too high a level of mechanising. Although teaching mechanics has greatly improved in the last few decades, I am not sure whether developers that follow the structuralist trend are really aware of the great measure of implicit (and therefore to themselves self-evident) horizontal mechanisation hidden in the contexts they offer to the learners.

Another feature by which the use of contexts in such mathematics instruction as I have in mind differs from most uses in mechanics instruction (in particular of the structuralist trend) is that of wealthy versus narrow-gauge contexts. A wealth of context which would admit and invite isomorphic mechanisations would be a breeding-ground for vertical mechanisation. This requires, of course, the wealth indeed offered with the view on vertical abstraction and generalisation, which however, is virtually absent in courses of the empiristic trend.

Theme, a modern looking didactical notion, has been interpreted by educational developers of the structuralist peruasion as a strait-jacket, rather than as a context. In mathematics education I have signalled long ago what I have called the *system constraint*: the fact that systematics is a characteristic feature of adult science, is judged to imply science taught according to the system, or even, the mere system taught, rather than the science organised by it. This is an old tradition which still survives in such titles (Chevallard 1985) as "La transposition didactique du savoir savant au savoir enseigné". At any rate it is contrary to the view on learning processes as guided reinvention.

DIDACTICAL PHENOMENOLOGY OF FORCE – THE STATIC ASPECT

Although I am well aware of the all but perfect state of my experience, I am going to venture an attempt at didactical phenomenology, in order to enforce my arguments. To prevent misunderstandings let me emphasise that the order in which I have presented the result need not at all determine the didactical sequence. Moreover the learning strands I have isolated in theory, should in practice be intertwined with each other.

Among the fundamental magnitudes in mechanics (and physics at all), and certainly on the level at which mechanics is taught, length and time no longer offer any problems; mass does, albeit moderately, as we will see later on. Derived magnitudes are not likely to offer big problems either. As such let me mention frequency, speed, even acceleration, but also density (which I will recommend as an access towards mass). What regards force, however, one has, in first instance, to contend with everyday- semantic troubles, which are a well-known linguistic feature; but even in the language of physics the meaning of "force" has been settled only after centuries long hesitations, and even now such forces as the centrifugal one are kept alive, albeit with the adjective "apparent". When dealing with the so-called elementary machines, Descartes took as force what we now name work, yet in the kinematic context, force was a function of velocity to Descartes as well as to Leibniz (though not the same for both of them). This kind of terminology still survived in the 19-th century – just remember the dispute around he "law of conservation of force" (which then meant energy)! Even in scientifically tainted everyday technical language, force still stands for a variety of magnitudes (which ones, strongly depends on the related vernacular). One may readily conclude that for the greater part, rather than at physics, the aforementioned tests on force aim at semantics.

To *measure* length, time, mass, and a number of derived magnitudes there exist more or less trivial instruments (even speedometers are now trivial, that is, self-convincing in a variety of situations. Force, however, looks different, but is it really so? If there is a difference, it is more deeply rooted, and a priori to measurement. Unlike length, time, mass, and even velocity which look directly accessible to the senses, force seems to require a more abstract approach. As a phenomenologic-didactical demand, it is natural to start instruction at what is directly observable (and to stick to it as long as useful and feasible). With regard to force the phenomena nearest at hand to meet this demand are *muscle experiences* (one's own or the less directly observable ones of others), although in most cases thought-experiments may supersede the concrete appeal to muscle force. If the experiments are illustrated by drawings, *arrows* are very likely to emerge as representatives of "forces" – an informal use of vectors which cannot be formalised numerically unless measurement of forces will have been discussed.

1. Two people in front of each other push each other by breasts, foreheads, hands, or thumbs to "measure their forces": opposite forces and equal as long as no one falters, but if someone does, it is the "weaker" force that gives way. (Among animals, goats are a good example.)
2. A tug-of-war: As long as no party moves, equality of opposite forces is granted; facts of moving establish an order relation on strength. In this example an intermediary is intercalated between the human points of attack: the tightly stretched rope. By "stretch" forces, geometrically as well as conceptually in line with the human forces, are *metaphorically* ascribed to

each of the parts of the rope: opposite, and as long as nothing and no one moves, equal. If one of the parties lets go the rope, the other falls, brought down by his own force. The same may happen to both of them when the rope breaks. Between this hands, a single person can stretch a cord and so transform muscular into stretching force.

3. Someone suppresses a resisting object or stretches it: a spring, the lid of a cooking-pan, a rubber band. Muscular force is called in against a metaphorical force of *resistance* acted out by an object.
4. Once more the rope, now vertically stretched by a hand at its upper end and at its lower by a weight, which is again metaphorically interpreted as a force to which the weight yields as soon as the rope is released or cut. However, the muscular force which thwarts gravity as strong as is needed to prevent a free fall, can also be exercised by supporting the weight by one's hand, for instance.
5. The same support can be provided by such things as a *table* on which the object is lying. Again a metaphoric upward force is attributed to the table, elicited by the weight and just strong enough to counter the gravity of the weight. For the first time in our examples, human force has been eliminated, and so the metaphor has lost its metaphorical status. But where does the force of the table come from? The question will be answered later on.
6. The fall of the weight at one end of the rope can also be prevented by a weight at the other end provided the rope is being led over a *pulley-wheel* – an "equal" weight, of course. This is again in all detail the tug-of-war, but now fought by gravity with itself, rather than by muscular force.
7. Another instrument to counter gravity by inverting is the *lever*. How come? We will return to this point.
8. A cord, fastened at both ends, hangs vertically. Its typical form is determined by pointwise acting forces which reveal themselves as soon as it is cut: two pieces which before cutting held each other, to wit, by opposite equal (though pointwise variable) forces.
9. The hanging cord is now taut by a weight. At the point where the weight is hung, one notices three forces which must offset each other: two of them stretching the cord, and as a third the gravity of the weight. Adding vectors is nowadays an early subject in mathematics. If arrows have already been used to represent forces, it is natural supplying them to the drawing, and that in such a way that the geometrical sum comes true. The objection that this is somewhat premature as long as measurement of forces has not been tackled, is hardly valid. This is a new situation if compared with the former simple pairs of opposite equal forces, but more complication is possible and needed.
10. An object, say a wood cube, has been sawn, horizontally, or vertically, or in any other way. This has changed the reactions on exterior forces, on pushing, lifting, pressing, and on gravity. This is explained by the change or disappearance of certain forces which acted before on the cut and can – partly or entirely – be restored by gluing. There are forces within a body to hold its parts together and keep them from each other; each part (one may also

say; each point) is acted on by forces in any direction which, in the state of rest, neutralise each other, for instance by being pairwise opposite and equal; and the same holds true of bodies that touch each other to form a compound body.

11. A plank is bending under a weight: The system of forces within the plank is being changed to have all the forces neutralise each other. How this happens can vaguely be explained by the examples of the hanging and the charged cord (in 9). In a way, the question in 8 of how one end of the lever "knows" what happens at the other end, can now also be answered.

INTERMEDIATE REMARKS

In the above sequence I have tried to avoid what I have called an access on methodologically too high a level. Common sense, rather than being exorcised, is taken seriously. Common sense is to start the learning process by which it reforms and educates itself. So, rather than mechanics, the learners are taught *mechanising*. Forces, unless experienced personally, are doomed to remain phantoms. One of the textbooks I have in mind deals with forces through the theme "Traffic". A strait-jacket theme, that is the way it is, where muscular force is mentioned just once: when the learner is warned to distinguish the forces exerted on a car by engine and brake from the forces of stepping on the pedals, while no attention is paid to their interrelatedness. The bicycle is dealt with unrelated to human force, and walking is not worth mentioning. Socio- psychologists would call it estrangement.

There has been much philosophised on the epistemic state of the terms, concepts, rules, laws of natural sciences. The most reasonable idea about it, although familiar to scientists, has, being too simple, found little favour in the eyes of philosophers. It says that scientists try to understand phenomena by splitting them up into two parts, *ideal case* and *perturbation* – take as examples the ideal uniform inertial motion or the ideal uniformly accelerated free fall on the one hand and perturbation by resistance on the other! All the same, there are also ideal perturbations which in turn may be perturbed (resistance of the air, for instance, by turbulence), and so it does on, by feed-back or to perturbations of ever higher order. Mechanics is full of such ideals: statics, rigid bodies, point masses, line segments that represent ropes, massless beams, and so on. All of geometry, if interpreted as a theory of the space we live in, is such a – readily accepted – ideal case, corrected for perturbations by the surveyor. Although in mechanics, judging by the troublesome law of inertia, the borderline between ideal and perturbations is less easy to draw than it is in geometry, the "system" confronts the learner bluntly with neatly prepared ideal cases, even when the context suggests – unconscious or conscious – idealising to be left to the learner.

Systematists may be frightened by my reckless ignoring of the boundaries of statics. Yet the mechanical behaviour of solid bodies cannot be understood unless one is allowed to part with the idea of rigid body. Let us ask, indeed, as an anticipation, how the path of a bouncing billiard ball can be reconciled with the law of inertia as long as bodies have to be rigid at any price. Descartes, who believed in rigid bodies, proclaimed the conservation of absolute impulse a law: except the first, all his collision rules are trivially wrong. Leibniz, however, who was the first to use the word "elastrum" in this context (though the word itself is a bit older), discovered the law of invariance of the impulse vector thanks to his understanding of *elasticity* (in generality he even surpassed Huygens), as well as the invariance of "force" (in the sense of mechanical energy).

DIDACTICAL PHENOMENOLOGY OF MEASUREMENT

For measuring physical magnitudes three things are constitutive in essence:

- a concept of *equivalence*, in order to assign the same measure to equivalent objects,
- a way to *compound objects* which extends to the addition of their measures,
- a unit measure.

Here I have restricted myself to "extensive" magnitudes. There is no physical reality involved with the addition of densities, frequencies, decibels, temperatures (*difference* of temperatures being meaningful), while adding with the view on averaging is a merely mathematical operation. Measure values can be *scalar* (such as length, volume, duration, mass) or *vectorial* (such as path, velocity, force, course of time); measuring units are scalars anyway.

With a reference to my 1983, Chapters 1 and 13, I skip length, area and time, while volume will be dealt with only in connection with weight and mass. Weight is first experienced as a force; a pair of scales with its weights looks like a museum piece today. Some use the balance in initial arithmetical instruction: a weighing beam with equidistant and numbered hooks at both sides at which (equally heavy) plates can be hung; its equilibrium is to illustrate addition and multiplication sums. But in everyday weighing, reading the hand on a scale has already been superseded by digital electronic devices – who does still think of the actually weighing part within the instrument?

The symmetry of the balance exhibits the equivalence relation which is constitutive for weight, directly or indirectly (via weights); compounding is also a concrete feature, and the authoritative datum of the scale division on the spring balance can be checked by "weighing" weights.

The constitutive operation of compounding bodies is the same for weight and volume. So the linear relation between both in one and the same "species" looks obvious. Numerically the relation is described by means of the "specific weight", a species-bound character, indeed. The fact that one object is heavier than another in ratio, has been explained from ancient times by a tighter packing of matter. Crucial experiences about this, such as compressing something and packing bags, deserve to be made conscious, in order to pave the way towards "density" and "mass". To define mass as "volume times density" looks like a vicious circle, which however, is not that vicious if it helps learners get an inkling of what is mass. The true problem with mass is the apparent lack of a constitutive equivalence relation (while the constitutive operation of compounding is just the same as for volume and weight). It is a point to discuss later on, after the intervention of kinematics.

Yet right now one can tackle the delicate problem of the relationship between weight and mass. The crucial point is the insight that, unlike the scalar mass, weight is a vector, to wit a force. To start the discussion let me recall a story from my 1977, which reports a sequence of experiments made by a $5\frac{1}{2}$ year old boy. After many weighings with a small spring balance, it happened by chance and unintentionally that he pulled its hook. He was greatly surprised when the small bolt indicated a weight again, and repeated the trick endlessly, pulling in various directions, and aiming at various "weights". I did not immediately recognise the import of this discovery. Later I returned to the problem, asking who in the case of the freely hanging spring balance pulled the hook. It was a suggestive question, answered by myself too suggestively: "the earth, and that is called gravity". It is a pity that at the same time I did not continue with weighing on an inclined plane.

With the view on the age and life experiences of learners we have here in view, let me now invert the perspective: muscular experiences as a source of the insight that the thing measured by a spring balance is a force. Of course other sources are not to be excluded, for instance, the surprising effect of a breath directed on an electronic letter-balance. Nowadays one can get newton-gauged spring instruments. So the pupils can measure quite early the force (in newton) of a mass of 1 kg and, in order to instill the idea that weight depends on more and other parameters than mass, it can pay to have the weighing also done with masses on inclined planes. Subsequently the teacher can tell what happens at the equator, at the poles and on the moon. Old-fashioned pairs of scales are to *compare* masses; spring balances, if used to measure weights, have to be *regauged* wherever they are used.

What is the rationale of the units kg and N? The same question can be asked about m and s, but up to now we have kept silent about the "how and why" of the units. While this should not be a problem with the well-known m, s and kg, I would guess that N urgently asks for a motivation, which of course, must be historical, although it can hardly be given satisfactorily before the law $F = ma$ has been dealt with. In due course we will return to this point. In the mean time and for later use, let us remember the numerical *linear interdependence of mass and weight*.

After scalar measurements, but without prejudice to the didactical order, I will now discuss vectorial ones. I am starting with magnitudes that are geometrical, rather than mechanical,

path, velocity, acceleration,

while, as an example, briefly following de Lange 1980: A plane is flying on compass from A to B. Then:

flight path = compass path + wind path,

flight velocity = motor velocity + wind velocity,

flight acceleration = motor acceleration + wind acceleration.

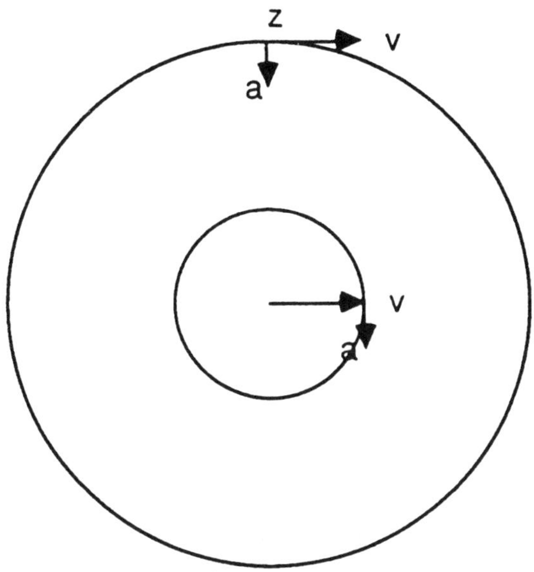

Fig. 1.

As usual, instantaneous velocities are preceded by average ones; in the long run, curvilinear paths are admitted. Translation is the constitutive equivalence relation for the various magnitudes; dragging along is the principle of compounding.

Before passing to measurement of forces, I add two remarks on the above more geometrical magnitudes: First, path, velocity and acceleration vectors are linearly dependent for any rectilinear motion. Second, for a uniform circular movement, the acceleration vector points to the centre. At a glance this follows from the drawing (Figure 1).

Indeed, if the velocity vector v is shifted from z to the centre, its arrowhead describes, simultaneously with z, a circle, the tangent vector a of which – the acceleration – is orthogonal to v, thus directed against the radius vector. (The second remark can be generalised as to apply to any uniform motion along a plane or spatial curve: the acceleration vector shows along the normal.)

Unlike the former, rather geometrical, vectorial magnitudes, force is a preeminently mechanical one. Force measurement was touched upon when "mass versus weight" was discussed. Besides rough estimates by means of muscle experience, the important principles of measurement in school mechanics are gravity and elasticity. Both are concretised by instruments, and their equivalence can be verified by regauging. It should be added that the measuring instruments which depend on gravity can be made fit for the measurement of forces in any direction by such tools as are pulleys and inclined planes.

The equivalence relation which is constitutive for force measurement is translation along the line of support: the point where one pulls the cord does not matter. Since force measurement involves more than simple geometry, it must be corroborated why vectorial addition accounts constitutively for compounding forces in one point. This can be done by a classical instrument: three pulley wheels mounted upon a metal circle and three weights acting, via chords over the wheels, on a point that must stay in the centre of the circle.

DIDACTICAL PHENOMENOLOGY OF FORCE – THE KINEMATIC ASPECT

What I have called "reckless ignoring of the boundaries of statics" is now rewarded by a smooth transition from the static to the kinematic aspect of force, which are usually separated from each other by a broad divide.

In my former series of examples a non-compensated force has been answered time and again by a *motion in the direction of the force*. More force means more motion – it is obvious, but one should make more precise what "more motion" means.

One can distinguish two versions of force in the examples: instantaneous force such as push and pull, and a force that keeps acting such as gravity and stretch; muscular force can it have both ways. When an object is pushed or thrown, does the (muscular) force that moved it, keep acting as long as the object moves, that is, at a distance, and how can this be possible? For two millennia this was a problem to struggle with. The historical solution was the *law of inertia*, although this final answer had already been implicit to the *principle of relativity* invoked every now and then by Galilei and Huygens: that the kinematic phenomena are the same on a smoothly sailing ship as on the wall; that on both one can, in the same way, drop and throw up a ball, play tennis and billiard, and balance an object. I would not like to miss this principle in instruction, if not as a source, then as a counterpart to the law of inertia. As in mathematics it pays to approach a goal from many sides; and certainly a goal like the law $F = ma$.

I know that in physics this view is not shared by most didacticians. In one of the textbooks I consulted the law of inertia is derived from a more or less dogmatic $F = ma$ by the reasoning that because of $m \neq 0$, the absence of force, thus $F = 0$, implies a vanishing acceleration a, thus uniform motion. Behold the physical law of inertia derived from a mathematical formula! A phenomenological road from the physical law of inertia to the mathematical formula $F = ma$? Forget it!

In the examples stemming from statics any extra force, however it came about, was answered by a motion. Rest is, rather than the opposite, a limit case of motion, a non-privileged state of motion – this is the principle of relativity. What matters didactically is to get across that force expresses itself by *changing* the state of motion. The instantaneous state of motion, that is. But what does this mean? In ancient times one did not know how to cope with it. Zeno's paradox of the flying arrow that cannot be on two places simultaneously and thus refutes the possibility of motion, preposterously fought as it was, kept on working for millennia. The breakthrough was due to the insight that for a mechanical description, besides the place, the velocity vector is required.

Force expresses itself by changing the state of motion, to wit, in the direction of the force, and this is called acceleration. For the state of rest this looks obvious, for a falling object it is no less so, but a particular impressive example is the uniform circular motion which, for instance, is kept going by a cord from the centre to the object swung around, thus by the centre-directed tightening force, in line with the muscular force of the one who swings.

As in the static case, I would start the learning process towards force in the kinematic case by means of muscular experiences as there are

push, pull, take off, thrust, tug, move, brake, strain, shrink.

We have already met the first two; they can change the state of motion directly or by an intermediary. The take off as a force is particularly revealing: from lying to sitting, from sitting to standing, from standing to running, leaping and jumping, but the most revealing, the jump from a springboard – the stepping stone is too obstinate and the trampoline is more involved. One does something to an object, which first yields, up to a limit, in order then to do something back. How much force to put depends on the kind of object, from a soft cushion to a springy plank, from slippery ice to rough gravel; in the first instance or as an intermediary the acted on object can be a sole or a glove or a jumping-pole or a prick to move a sled or a stake to move a boat – in the last two cases the object moved is not oneself but one's carrier. The football boot and the racket which make the ball rebound are examples where the *object* "takes off". The billiard ball takes off the cushion – now it is no more muscular force but the combined action of elastic forces as known from the statics of force.

Brake: braking one's own motion by pushing back oneself or an object in front (think of the trampoline!) – muscular experience which in a manifold ways can metaphorically be ascribed to objects. What one's muscles experience in stretching or shrinking oneself or a rubber string or a spring, in pressing or releasing, are again forces that keep on acting, among which is also gravity, in the static approach tested by muscle force. All this can be repeated kinematically: the yo-yo and the ball tossed and caught.

An object falls the longer the faster – this is qualitatively verified, with no measuring needed. Continuously the speed changes, and so does the state of motion – something that is only possible thanks to a force that keeps acting, gravity. Wherever the falling object comes, it is equally heavy, the force that changes its state is constant all the way. So it falls with a constant acceleration. No more is needed to derive the law of free fall (and Galilei knew no more).

According to experience it costs more force to get an object on the move than to have it moving, more force to change the motion than to keep it (the obstinate supermarket trolley!). One blow and the ball moves, on asphalt farther than on turf, even farther on ice and in the "ideal case" infinitely far, straight away and relentlessly. The ideal case is called *law of inertia*, the perturbation, *friction or resistance*. Even under the most favourable circumstances the transition from rest to motion requires a threshold of force but so does any change of state.

Although mechanics is concerned with the ideal case, one cannot now avoid to mention perturbation. The question to be answered is whether motion gets lost by perturbations, or where it goes. The question is eventually answered by "conservation of impulse", but it can early be made plausible that when motion looks lost, it has been yielded to the resisting medium. As early at Stone Age man knew how to produce heat, and wherever heat shows up it indicates internal motion, from mercury to steam-engine. Answering gravity, objects, out of a moving train or flying plane, do not fall vertically, but take along with them the velocity of their carrier, which then is changed by gravity, orthogonal to the original one (watch your step when you leave the train before in stops!). The object swung around flies off with its instantaneous velocity, that is, along the tangent, the same way a speck of dust comes loose from a turning wheel. (But at the turning Earth even the birds high in the air remain attached, thanks to gravity, as do you, although it makes a small difference how high you are jumping. A turning wheel keeps turning uniformly – though cited by Newton it is not the best fitting example to square with his law since there is force involved, pulling along the spokes towards the centre, which prevents all that is well enough attached to the periphery from flying away tangentially, thus again forces orthogonal to the instantaneous motion and fit to change it.

Still a question regarding the sled and the boat moved by a pole: how comes the inclined pole moves the carrier *horizontally*? Horizontal is the only way it can happen, indeed. The vertical component disappears in the ground. The opposite happens to the billiard ball at the cushion where the orthogonal component disappears and is even inverted while the ball continues its motion parallel to the cushion (not considering possible spin).

$F = ma.$

The formula can be brought up as a mere postulate, but it can also be derived from the supposed

a. proportionality of mass and weight,
b. vector addition of acceleration,
c. vector addition of forces,
d. vector dependency of force and effected acceleration.

From (d) it follows that the force vector is a scalar multiple (say the β-ple) of the acceleration vector. Thanks to (b) and (c) it is clear at a glance that β is the same for all, however directed, forces, and according to (a) it must be proportional to m. Thus F and ma are equal up to a universal factor which can be rid off by norming, that is by an appropriate choice of measuring units (Figure 2).

ONCE MORE: MASS

To satisfy the requirements for measurement, we still lacked an equivalence relation for mass. After the law $F = ma$ it is obvious how to formulate it: Masses are judged equivalent if they react on equivalent forces with equivalent accelerations.

Of course, here we are concerned with mechanical force, moreover with a view on the derivation of $F = ma$, restricted to point masses.

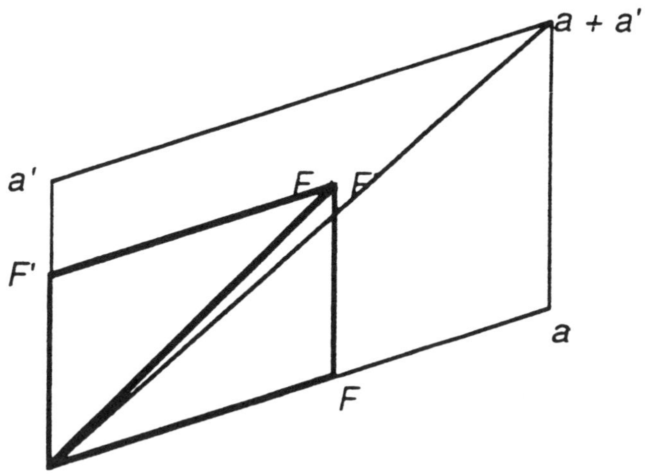

Fig. 2.

MEASURE UNITS

The French revolution brought a reorganisation of the system of measures. The unit of length was defined to make the pole-equator distance 10000 km. Since the required meridian arc measurement was flawed, the *standard metre* came out a bit too short and hence the earth quadrant a bit longer than intended, but the metre was maintained as a standard. Area and volume measures were linked to it, and it was settled that (under certain conditions) 1 dm^3 water should weigh 1 kg. With regard to time the astronomic day with its usual subdivisions, though a departure from the decimal system, was maintained. As precision of measurement improved, units were repeatedly redefined; at present, cm for instance, by so and so many times the wave length of some spectral line, and s, similarly, by a vibration period.

In theory all physical units are based on those of length, mass and time, and that in a way that no superfluous constants show up in a number of basic relations, for instance, cm s^{-1} and cm s^{-2} as units of velocity and acceleration. 1 kg, defined originally as unit of weight (thus of force) was in the long run interpreted as unit of mass. To meet the troublesome conceptual distinction between kg-weight and kg-mass one redefined the unit force to cause the unit acceleration at the unit mass and gave it the name newton (N), by which

$$F = ma.$$

is granted with no proportionality constant involved, thus

$$1\text{N} = 1 \text{ g } 1 \text{ cm s}^{-2}.$$

In practice (and certainly in everyday weighing), the kg-force has been maintained, which at our place is 9.81 N.

PREPOSTEROUS FORCES

The Centrifugal Force

At a sudden stop of the bus the upper part of your body dashes forward. Who pulled it? Nobody. Instead, the bus pulled your bottom at which the upper part of your body is so flexibly attached that the braking force did not get an immediate grip on it; so lack of force was interpreted as force. Or, sitting at the left, when the bus takes a right bent, you feel the bus press your side. Does it? The centripetal force acting on the bus that changes the state of motion of the bus yet not yours relative to the bus is interpreted as a centrifugal force acting on yourself. The cord that swings round an object feels as though taut by a centrifugal force. This is what the object is said to feel, while the one who swings keeps the object centripetally from flying away. When the merry-go-round goes turning the hanging boats are striving upwards (the only way to fly) until equilibrium is reacted between gravity and forces pulling towards the axis, which felt centrifugal by people in the boat.

It is, of course, a matter of definition what one wants force to mean. If we wish that pupils stick to our definition it is a good policy to confront them explicitly with all terminology at variance: magnitudes that once (and certainly not without reason) have claimed the name of force and still are associated with it instinctively. Prevention is better than cure: making conscious should take over from shots which promise, but not always grant, immunity.

Velocity and Impulse

The Meteorological Institute gives wind-force in metres per second; the News-Report does it in kilometres per hour, that is, 36 km h^{-1} instead of 10 m s^{-1}, obviously because this sounds better to drivers and cyclists. But how can a cyclist, with his 20 km h^{-1}, manage to pedal against a wind of double his speed? The same way as a calm walker resists a galloping child. What, besides speed, counts in collisions is mass. The product of mass and vectorial velocity is called impulse. Why the product?

For the impulse a conservation law holds true; it rules *elastic collision*. In absence of external forces the vector sum of the impulses of a system remains constant. When two masses approach each other head-on, with (absolute) velocities in the inverse proportion of their masses, the sum vanishes before and hence after the collision; which in the case of an elastic collision signifies that after the shock both return the same way they arrived. As a matter of fact the law of conservation of impulse is nothing but the law of inertia applied to the centre of gravity (or rather, of mass) of the system, where one may think its masses united, indeed.

These are well-known facts but in this form they do not yet answer the question of "why the product?" I have not yet succeeded to formalise by ideas at this point.

Historically impulse is known as the cartesian concept of force – absolute impulse[9], that is, for which Descartes (wrongly) claimed conservation in collisions. Still today impulse is willingly and certainly not without reason intuited as a force, an inner force which is, as it were, personally experienced when running and cycling, then ascribed more or less metaphorically to the car or train in which one is seated, and finally to all that flashes by. Not without reason, I said, and this justifies the didactical demand that it is explicitly dealt with in instruction: that in our terminology, force as far as active, is not something interior, hoarded within its bearer (and as such conserved), but rather something exterior and extra that may change a given state.

Work and Living Force

Originally Descartes had another idea of force: force (in our sense) times path, nowadays called work[9]. It turned up when he discussed the elementary machines: the Golden Rule of yore, kind of conservation law, stating that even machines do not allow one so save "force"; the tackle, for instance, only spreads out the work as the putative gain on the one hand is paid back on the other hand by correspondingly greater lengths one has to go.

With this concept of force Leibniz fought the cartesian one (impulse), albeit from another starting point. A mass point, when rolling off an inclined plane, acquires a velocity that allows it to climb another inclined plane up to the same height – kind of conservation law. The "force" required to lift a mass m against gravity to the height h is mgh. And the velocity v with which it returns when it falls back? If t is the time needed for the fall, one has

$$h = \tfrac{1}{2}gt^2,$$

$$v = gt,$$

thus after elimination of t,

$$mgh = \tfrac{1}{2}mv^2,$$

at the right the *living force*, at present called *kinetic energy*, the counterpart of the *potential energy* at the left. Their sum remains unchanged during the entire fall. For Leibniz this was the law of conservation of force, which in the 19th century was extended beyond "ideal" mechanics to include the theory of heat. Not unlike impulse, *energy* is intuitively felt as force, and again for good reasons – a force hoarded and spent by turns, as it were – and again this is a didactical opportunity for disclosures.

However strange it looks, energy, among all mechanical notions and after all the trouble physicists had had with it in the past, now, in our industrialised world, seems to be the most concrete notion – concretised, among others, by gas and electricity bills. So it is not to wondered at that in textbooks the chapter Energy may be one of the best. But this only reinforces my view that one should immediately, that is, nearby the source of future misapprehensions, confront energy with *power*. The confusion "kilowatt/kilowatt-hour" is a well-known and not merely terminological phenomenon, or so I guess. The relationship between both of them deserves better than only being posited as mathematical relation; it deserves to be broadened and deepened at the very beginning, with the electricity bill; what is the fixed amount be charged for? The relationship between energy and power has its counterpart wherever variable production takes place on a constant basis: fixed versus use-dependent charges. The scope of this distinction extends so far outside physics that instruction should certainly account for it.

Resistance and Friction

These are preposterous in another sense than the preceding ones: forces, though by definition and nature stemming, nevertheless indispensable to moving ahead. Since I have already paid enough attention to these forces, I restrict myself now to emphasising once more the didactical need for clarity around this seeming paradox. In no way do rigid bodies suffice for a solution, which moreover, has to be sought for first in the domain of one's body. I have looked for a neutral term for the forces at issue, which might help to unravel the paradox. Would "give-and-take force" do it?

Action and Reaction

Never have you heard me pronounce here this pair of words. However interesting it might be from a historical standpoint, I simply do not know how to use it in instruction.

CONCLUSION

I have moved in a domain I have hardly explored. It was predestined to remain an attempt, in no way definitive or conclusive. What I could offer was an illustration of what ensues from a didactical principle: If it is true – and who would deny it? – that in mechanics instruction body experiences interfere inconveniently with scientific ideas, then it is of paramount importance, instead of suppressing them, to have the learning processes started just there and the learner, under guidance, transform them in what we consider scientific.

NOTES

[1] For instance: Freudenthal 1977.

[2] Freudenthal 1990.

[3] Van Genderen 1989.

[4] PLON 1983, 1988, 1989.

[5] More thoroughly: Freudenthal 1990b.

[6] Segre 1979.

[7] Treffers 1987.

[8] Freudenthal 1983.

[9] Freudenthal 1985, 1986.

REFERENCES

Chevallard, I.: 1985, *La transposition didactique du savoir savant au savoir enseigné*, La Pensée Sauvage, Grenoble.

Freudenthal, H.: 1977, 'Bastiaan's experiments on Archimedes' principle', *Educational Studies in Mathematics* **8**, 3–16.

Freudenthal, H.: 1983, *Didactical Phenomenology of Mathematical Structures*, D. Reidel Publ. Co., Dordrecht.

Freudenthal, H.: 1985, 'Die Gesetze des elastischen Stoßes bei Leibniz', *Boethius* XII: Mathemata, Festschrift für H. Gericke, 427–439.

Freudenthal, H.: 1986, 'Autour du principe de continuité de Leibniz', *Studia Leibnitiana*, Sonderheft **14**, 157–168.

Freudenthal, H.: 1990a, 'Mijmeringen over mechanica-onderwijs', *Tijdschrift voor Didactiek der β-wetenschappen* **8**, 222–248.

Freudenthal, H.: 1990b, *Revisiting Mathematics Education*, Kluwer, Dordrecht.

Genderen, D. van: 1989, *Mechanica-onderwijs in beweging – een onderzoek naar veranderingen in de didactiek van de mechanica in het voortgezet onderwijs*. Proefschrift Utrecht.

Lange, J. de: 1980, *Vlieg er eens in*, IOWO RUU, Utrecht.

Mach, E.: 1912, *Die Mechanik in ihrer Entwicklung*, Brockhaus, Leipzig.

PLON: 1983, *Verkeer.* Vakgroep Natuurkunde-Didactiek RUU, Utrecht.

PLON-vereniging: 1988, *Natuurkunde in thema's 2a*, Zeist: nib.

PLON-vereniging: 1989, *Natuurkunde in thema's 2b*, Zeist: nib.

Segre B. et al.: 1979, eds. Congresso Internazionale, *L'insegnamento integrato delle scienze nella scuola primaria*. Roma, 7–15 gennaio 1976, Accademia N. Lincei 326.

Treffers, A.: 1987, *Three Dimensions*, D. Reidel Publ. Co., Dordrecht.

Freudenthal Institute
Utrecht University
Utrecht
The Netherlands

A. TREFFERS

WISKOBAS AND FREUDENTHAL REALISTIC MATHEMATICS EDUCATION[1]

ABSTRACT. Freudenthal was the founder of so-called realistic mathematics education. In it reality does not only serve as application area but also as the source for learning. We take a newspaper article as an example of a rich context problem. It contains elements from all of the important areas of learning for mathematics education in primary school (grades 1 through 6). In the following these topics of learning are discussed. Historical comments will be indicated each time, in particular relating to the influence of Freudenthal. Freudenthal laid the foundation for this didactical realism and determined the development of various learning strands, but more indirectly than directly because he himself did not design or outline themes and learning strands. Attention is devoted briefly to the integral educational development of Wiskobas and Freudenthal's contribution to it, especially in terms of developmental research. The summary, in conclusion, is given by way of a problem that was posed by Freudenthal – his last one.

INTRODUCTION

Freudenthal is strongly associated with Wiskobas[2]. This is rightfully and wrongfully so. Wrongfully so because Wiskobas originated in 1968 without the involvement of Freudenthal and continued after 1981 without his direct participation. Rightfully so because of the lasting influence of his range of ideas on the work of Wiskobas. And not only on Wiskobas, and therefore on primary education, but equally through Wiskivon on secondary education (grades 7–12), which will not be described in this presentation[3].

It was Freudenthal who in 1971 – upon the founding of the IOWO – put Wiskobas on the track of realistic mathematics education, away from the trodden path of traditional arithmetic, and away from the up and coming New Math. New in Freudenthal's views was not only that he wanted to incorporate everyday reality emphatically in mathematics education, but especially also his fundamental idea to let that rich context of reality serve as a source for learning mathematics.

In the initial period it was predominately the rich contexts in the form of topics, projects, locations, stories and clippings which determined the appearance of Wiskobas. Later it was more the new learning strands which drew the attention.

In the following there is first an example of one of those themes for the teacher, together with some historical comments about the place of such problem-situations in realistic mathematics education in primary school (grades 1–6) in the period 1970–1990, and Freudenthal's contribution towards the development of these themes. Thereafter the latter is done for the learning strands of the Wiskobas

Educational Studies in Mathematics **25**: 89–108, 1993.

Fig. 1. Photo: EPA.

program and in conclusion for educational development as a whole. We start by accentuating the description of the content of the Wiskobas program. Later it is especially Freudenthal's influence as a member of the Wiskobas team which receives attention.

1. RICH THEME: VAN GOGH IN THE FIELD

1.1. Van Gogh Problem

From the outset of the Wiskobas project at the IOWO Freudenthal stimulated 'mathematics fraught with relations', or as it was later called 'mathematics in rich contexts' – to which the influence of Decroly was not foreign (Freudenthal, 1973). From the available categories of rich themes a newspaper clipping is chosen here, intended for the (initial) teacher training. To be precise the example consists of two newspaper clippings, both about a rather unusual reproduction of one of Van Gogh's 'Vase with Sunflowers' (see Figure 1).

> It took Andrew Scott, a horticulturist in Scotland, ten days to make a reproduction of Vincent van Gogh's 'Sunflowers'. With the assistance of twenty volunteers he copied the painting in a Scottish wheat field, using 250,000 colourful garden plants. The botanical 'canvas' is 14 square kilometres in size and "very beautiful", Scott concludes after having inspected the scene from a helicopter. (Trouw 17–8–'91).

In another newspaper one could read:

> Scott used various kinds of plants and shrubs to portray a picture of over 10.000 square meters. (Volkskrant 17–8–'91).

On the basis of these clippings we asked the teachers the following question:
> Which report is the most accurate, Trouw or Volkskrant? Explain your answer as convincingly as possibly.

When this problem had been solved and discussed the next question was raised:
> How do you suppose this discrepancy in sizes occurred?

It was added that the newspapers were given the area in square yards: 12820 square yards (De Moor and Streefland, 1991). More explicitly:
> Could the area of 'over 10,000 square meters' be correct?
> How do you explain the error of the 14 square kilometres?

1.2. *Brief Comment*

In determining which area seems the most plausible, there are first of all the arguments of geometry which the teachers raise. In making estimations of length and width tractor tracks and the height of the trees were considered: how wide is a track and how high a tree, roughly speaking? These sizes as such are the first topic of discussion and reflection. All kinds of reference points are brought up. One teacher follows up with an interesting method: the 'projection' of the side-lines of the painting to the road with trees showing at the top of the photo (Figure 2).

It is obvious from looking at the trees, that this 'canvas' cannot be 3 or 4 kilometres wide, but some one hundred meters at most.

Then there is also the computational argument in answering the first question with regard to the density of the plants in the painting: 250,000 plants on 14 square kilometres or on 10,000 square meters? In the latter case there would then be 25 plants on 1 square meter – that would seem reasonable, after some sizing up. But 250,000 plants on 14,000,000 square meters is 25 plants on 1400 square meters or 1 on 56 square meters, so 1 plant on a rectangle of 7 by 8 meters, the floor of our classroom, and of course that is too far apart.

Moments of discussion are introduced when making an inventory of the various solutions. But the result of one and other is clear: 10,000 square meters seems by far the most realistic size. It could also very well concern a rectangle of 60 by 70 meters, as reported on the news on television the day before! (Length and width do not differ that much, but the area does. How is that possible?)

Then the correct area is given – 12,820 square yards. We take another look at those 10,000 square meters and 14 square kilometres and do so in that order.

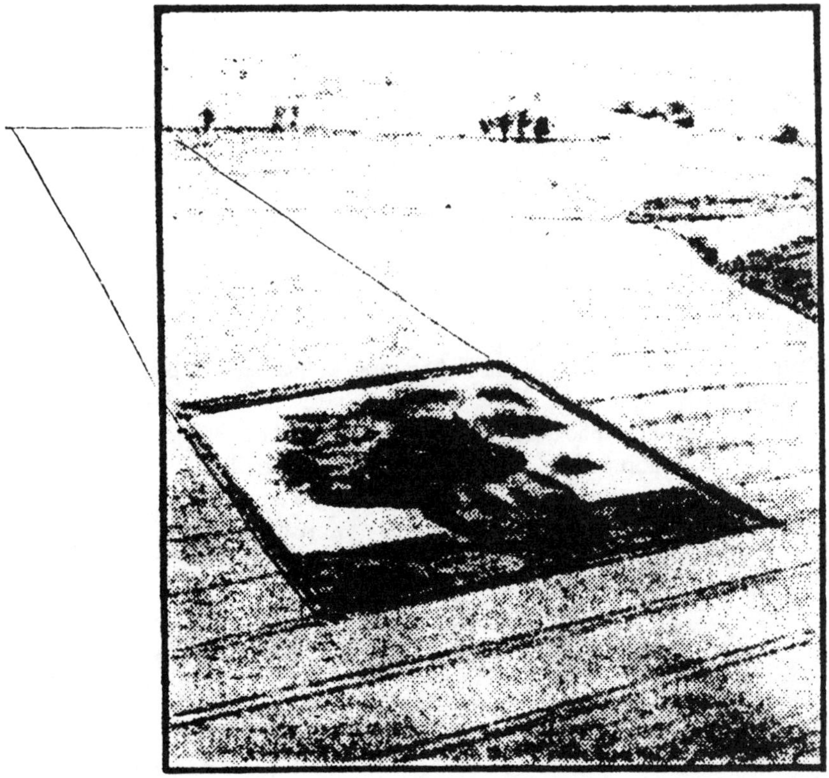

Fig. 2.

When converted does 12,820 square yards correspond with somewhere close to 10,000 square meters? One yard is 91.44 cm, let us say 0.9 meters.

There is disagreement: one group feels that the answer should have been well over 11,000 square meters ($0.9 \times 12820 = 115380$) or even 11,500 square meters; someone else prefers close to 12,000 square meters, and there are those who can agree to the figure of some 10,000 square meters (0.8×12820).

Of course the issue of rounding off is discussed at length. But particularly the main issue of whether to multiply by 0.9 or by 0.8 (or divide).

A square of 10 by 10 yards can serve as a model.

This is then converted to a square of 9 by 9 meters. Therefore 100 square yards corresponds with 81, let us say 80 square meters.

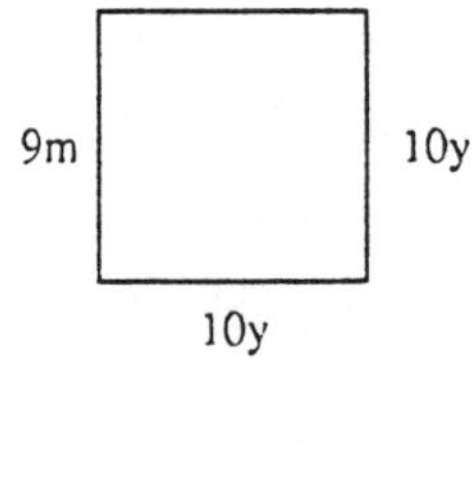

If 100 becomes 80 than 12820 therefore becomes ... The double number line is used:

100	12820
80	?

Or perhaps better the ratio table, because of the possible confusion of 'square' with 'length' on the number line:

100		10.000		12820
80		800		?

Different methods of calculation are possible; for example via 10,000 and correspondingly 8,000, or abbreviated via 130 times as much at the top, therefore also 130 times 80 at the bottom, or via 8/10 times 12820 – some 10,000 square meters does appear to be correct!

Actually this problem is best posed separately and the solutions inventoried and discussed one by one. But even more important is to relate the conversion of square yards to square meters to the earlier conversion of square kilometres to square meters, and to draw attention to the matter of the zeros and the linear connection between kilometres and meters, and the conversion of measures of volume.

In conclusion the unravelling of the matter of the 14 square kilometre.

The teachers are helped on their way: the first mistake also occurred here in the conversion of square yards to square meters via the factor 0.9. Who remembers what else went wrong? We latch on to a smart suggestion from one teacher: indeed 12820 was not multiplied by 0.9 but divided by it. Perhaps an understandable error, just like the first one, because the idea was that the result in square meters would have to be smaller than 12820, and becoming smaller means that you must divide, right?!

Next, the result of 12820: 0.9144 will probably be worked out on the calculator: 14020.122. The decimal point is easily overlooked. Now move the decimal point six positions to the left and we have 14.020122 square kilometres, rounded off to 14 square kilometres.

That is how it might have gone, but more likely the decimal point was placed three positions to the left (from meters to kilometres as it were) from 14020.122. Whatever the case, the error has taught us that division does not always give a smaller result. And the most important thing that was learned is perhaps, that in cases like these you must always examine the answer to see if it is plausible. By the way, one question remains unanswered: how did the Scottish arrive at 12820 yd. $2^2 \cdot 5 \cdot 641$?

At any rate it has become clear that many aspects of mathematics can be raised here in an integrated fashion. In the early days of Wiskobas, but also thereafter, theme-oriented examples like this, but also at the level of primary education, had an important informative and stimulating influence.

1.3. Historical Comments

Before the founding of the IOWO, Freudenthal wrote the following about mathematising:

'The globally structuring force, as we called it, should be lived through reality. Only this way can we teach mathematics fraught with relations, can we be sure that the student integrates the mathematics he has learned, and can we guarantee the applicability of mathematics'. (Freudenthal, published in 1973, p. 132).

It is evident that the introduction of rich themes corresponds with this basic idea, i.e. was strongly stimulated by it. The IOWO snapshots, published in Educational Studies in Mathematics on the occasion of 'H. Freudenthal's Retirement from the Directorship of IOWO' (1976) attest to this: 'An Island of Geometry', 'Ship Ahoy', 'Our Earth', 'Building a Bungalow', 'Gulliver', etc. Many themes were developed in the course of time. In his farewell speech Freudenthal even went so far as to predict that the integrative aspect of mathematics was dominant to such a degree that mathematics as a separate subject would cease to exist in time:

'What will mathematics education look like in 2000? The answer is simple. There will be no more mathematics education in 2000, it will have disappeared. There will be no more subject called mathematics, no math program, no math textbook to teach from' ... 'It is there to be lived and enjoyed, just as reading, writing, handicrafts, art, music, breathing, in integrated education'. (Freudenthal, 1977, p. 294).

After 1976 mathematising was distinguished in a horizontal and vertical component (Treffers, 1987) and also in the functions which rich context can accomplish in mathematising (de Lange, 1979, 1987). The integrative themes have a specific horizontal mathematising purpose, i.e.: they are primarily directed at the application, practice and relating of learned knowledge and skills in context situations. The model-type contexts on the other hand have primarily a vertical mathematising purpose, i.e. that they are in particular directed at the progression within the subject structure, therefore on the practising of knowledge, skills and concepts within learning strands (Gravemeijer *et al.*, 1990). By this distinction in horizontal and vertical mathematising, about which Freudenthal said "For a long time I have hesitated to accept this distinction" (1991), the importance of the integrative themes was brought back to realistic proportions: they are important, but overemphasis leads to empiristic mathematics education with too little attention for vertical mathematising. In an evaluation discussion about the Wiskobas project[4] in the IOWO period (1971–1981) Freudenthal himself came to the conclusion that Wiskobas spent too much energy in those days on the design of themes ('domains of reality disclosed to the learner in order to be 'horizontally' mathematised') and too little on model contexts ('to be 'vertically' mathematised') on which in the eighties the emphasis came to lie much more. Also the form and content of the theme changed in the eighties to some degree. Besides themes, projects, locations and stories there appeared the clippings. Their value was recognized by Freudenthal:

> Here is not a domain but a small piece that is cut out, although its paradigmatical value for mathematising and for acquiring a mathematical altitude may be enormous in comparison. (Freudenthal, 1991).

See 'Van Gogh in the field', we might add.

2. WITH A VIEW TO LEARNING STRANDS

In the Van Gogh problem there were elements from the main learning strands of the Wiskobas program:

- basic skills and column arithmetic
- ratio and fractions
- measurement and geometry.

In the following these learning strands will be regarded from the points touched on in 'Van Gogh' to acquire some insight into the specific characteristics of each. For each main learning strand some historical comments are made about their development in the period 1970–1990. In each instance there will be a short reference to the involvement of Freudenthal.

2.1. Basic Skills and Column Arithmetic

Mental arithmetic is efficient calculation for which both memorized knowledge as well as insight in arithmetic rules is necessary, with furthermore the particular 'faces' of numbers and the relations between them. The Van Gogh problem offers various possibilities for mental arithmetic in combination with estimated calculation, for example of $\frac{8}{10} \times 12500 = 8 \times 1250 = 4 \times 2500 = 2 \times 5000 = 10,000$ – the strategy of halving and doubling. The Wiskobas program leaves much room for mental arithmetic. Column arithmetic is not introduced until the beginning of grade 4. Until that time mental arithmetic has received every opportunity to develop. In the traditional math program column arithmetic (with adding and subtracting) started one year earlier, at the beginning of grade 3, and in consequence snowed mental arithmetic under completely. There are even countries (among them the USA) where column arithmetic to one hundred often already starts in grade 2. In consequence pupils have almost no opportunity to consider numbers on their value, something which is also expressed in the achievements (for a comparison see Treffers, 1991a). Estimation arithmetic has meaning when it concerns the rough determination of the result, the general assessment of the result in terms of magnitude, and if one must handle not precisely determined information. And that is exactly what was done in the Van Gogh problem, where experience information was used (misused), approximation, rounding off and (in)accuracies in determining the height of trees by roughly converting yards to meters and square yards to square meters and examining whether about 10,000 square meters is close... Although the Wiskobas program does devote much attention to estimation – it is both a goal and a means – this is not yet sufficiently

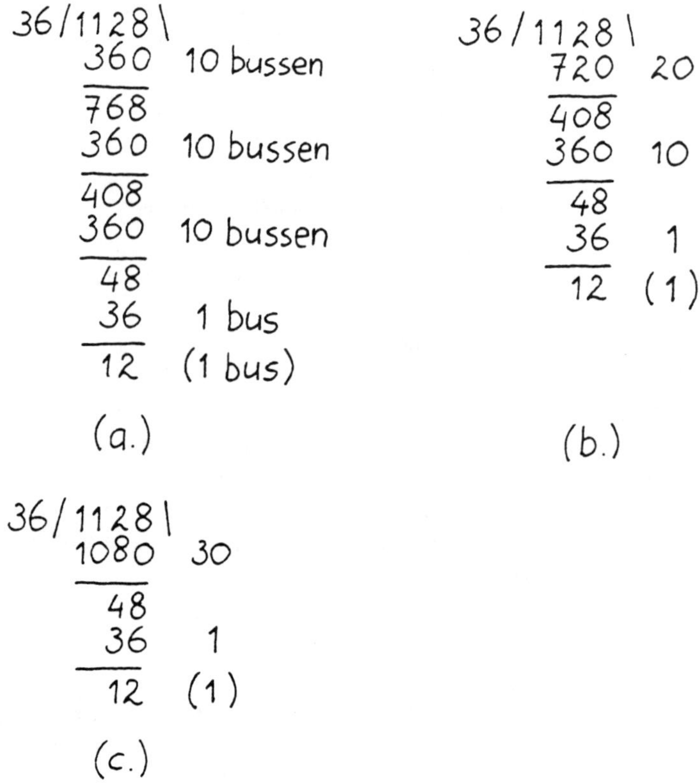

Fig. 3. Buses.

evident in educational practice in Holland, contrary to mental arithmetic that is practised sufficiently.

In conclusion, column arithmetic that in 'Van Gogh' could remain in the background by using the calculator. But this also applies in general: the importance of learning the number algorithms has been strongly reduced by the introduction of the calculator. In the Wiskobas program this is expressed in a special manner, namely in the connection between mental arithmetic and column arithmetic. Column arithmetic procedures are developed from stylized mental arithmetic. This occurs according to a gradual process of schematising and abbreviation. And the end level can differ (per group or per pupil). An example of how this works for division. The following transportation problem can for instance serve as point of departure for long division: 1128 fans must be transported in buses that seat 36.

How many buses will be needed?

This problem is solved at different levels in the course of the learning strand (Figure 3).

And these solutions illustrate the abc of the entire learning strand.

Something similar happens for multiplication. For example in a context problem in which '7×24' occurs: (1) via repeated counting, (2) abbreviated to the front-end method $7 \times 24 = 140 + 28 = 168$, and (3) the end-front standard algorithm that starts with the units.

Historical Comments

One might even go so far as to elevate the stylized front-end methods of mental arithmetic to (new) standard algorithm. The Wiskobas program, which serves as informal National Curriculum (Treffers, De Moor, 1990), emphatically points out this possibility. Currently it is not clear to what degree textbooks and teaching practice are prepared to go with this. There is clarity about the modest importance of column arithmetic: currently there is an obvious trend in The Netherlands to delay it (let us say until grade 4, as proposed by Wiskobas) and to restrict it in favour of mental arithmetic and estimation arithmetic. This shift did not take place until in the eighties however. During the IOWO-period (1971–1980) the mentioned parts of basic skills and arithmetic were still fairly unrelated: there was a separate algorithm program. And column arithmetic was introduced in grade 3 in those days.

No special didactics had been developed for the learning of the tables of multiplication or elementary mental arithmetic. That did not occur until the eighties for addition and subtraction (Treffers, 1991b), for multiplication and division (ter Heege, 1985), for elementary mental arithmetic with the aid of the empty number line on the basis of an idea by Whitney (1988), and for the combination method mental arithmetic-column arithmetic (Treffers and De Moor, 1990). Programs were developed for mental arithmetic for instance, and the writers of math methods also gave an important impulse. In short, too many people to mention together brought about a small revolution in the program development for basic skills and arithmetic.

In conclusion, a few remarks about Freudenthal's influence on the outlined development of basic skills and column arithmetic. Contrary to what one might expect, Freudenthal was hardly involved at all with this area of development. And as far as the above mentioned issue about the relation and shift in position of mental arithmetic and column arithmetic is concerned, one will find nothing on the subject in his last book (Freudenthal, 1991). He was not keen, so it became evident in the early stages of Wiskobas, on the fundamental importance of memorized knowledge of addition and multiplication tables. This did change somewhat under the influence of the Wiskobas group, but the fact remains that this area certainly did not have his special interest. It lay much more with the other major areas, which we will now review from the Van Gogh problem.

2.2. *Ratio and Fractions*

The first of the 'Van Gogh' problems concerns visual ratios. Numbers barely play a role in determining that 14 square kilometres cannot be correct: the practised eye sees immediately that this cannot be a rectangle of $3\frac{1}{2}$ by 4 kilometres, considering the trees, the tracks, the whole scene. Ratios also appear in the

Wiskobas program in similar fashion, therefore at first only as visual ratios from the observation world of drawings, maps, models and illustrations, whereby it must be determined whether the ratios are correct or have to be adjusted. Later in the program a number element is added. At issue then is estimation or determining the exact ratio relationship, whether ratios are equivalent, or else to compare ratio relationships, or determine the fourth proportional. In the 'Van Gogh' item for instance, visual ratios were concerned, to determine the plant density for the two given measurements and the conversion of (square) yards to (square) meters. But of course there are many more forms in which ratios can appear: mixing, borrowing, conversion, fair distribution, and the linear relating of measures such as unit-price, price-weight, road-time and so on, to which the mentioned questions about ratios can be applied. The dominant think and calculation models employed are the (empty) double number line and the ratio table (also consider the (estimated) conversion of 12820 square yards to square meters) which can both serve on many levels of schematising and abbreviation – of course also for the special ratios which are called percentages. Wiskobas has from the very start devoted much attention to this rich area.

Fractions do not play an important role in the Van Gogh problem. Only in the relationship between yards and meters do they appear, and then only in the form of decimal numbers. And working with them can even be avoided by size reduction or by working with squares of ten by ten. This immediately indicates one of the strategies with which to approach and explain the multiplication and division of decimal numbers, namely by changing measurement units with the help of the rectangular model or else trying to get rid of the decimal point. But fractions can also be used to solve ratio problems: ratios and fractions can be related to each other. Ratio (fair) division for example of pizzas, chocolate bars etc. can lead to breaking up and hence to fractions. Sharing of objects can achieve the same. Accurate measurement with a prescribed measure also. And thus there are many inroads to fractions which are used by Wiskobas in the first concept-forming phase. At issue is not so much operating with fractions in the more formal sense, but fair sharing, distributing and measuring and then the description of these activities and their results – the development of comprehension and language. The relationship with ratios is expressly maintained at this stage, partly to compare fractions and to determine the (in)equality thereof through ratios (Streefland, 1991a). In a next phase when operating with fractions is brought more to the fore, it is especially the measurement aspect that is employed. Keywords here are: mediating unit, named fractions, the rectangle model and the strip or the (empty) double number line (Streefland 1991a, b, and Treffers, 1991b). An example from the last in conclusion:

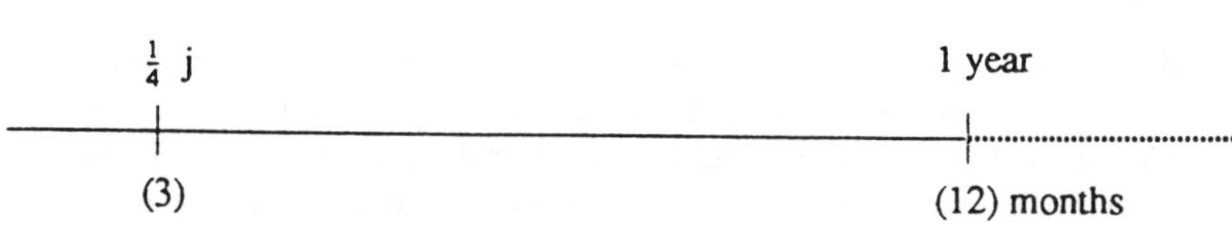

Problems such as $\frac{1}{3}y + \frac{1}{2}y$; $\frac{1}{3}y - \frac{1}{2}y$; $\frac{1}{3} \times \frac{1}{2}y$; $\frac{1}{3}y : \frac{1}{2}y$ can be solved via the mediating unit of month, therefore the underside of the double number line, but of course also directly with fractions on the upper side of the number line. Note the difference in notation between adding and multiplying: $\frac{1}{3}y \times \frac{1}{2}y$ is not possible but $\frac{1}{3} \times \frac{1}{2}y$ is right, i.e. that $\frac{1}{3}$ serves as operator. In the abstract notation $\frac{1}{3} + \frac{1}{2}$ and $\frac{1}{3} \times \frac{1}{2}$ one cannot read the different status of $\frac{1}{3}$ in these notations, as one can in the notations with letters. Later on in the course pupils must themselves try to find the right divisor to be able to work away the fractions by imagining a chocolate bar with a varied number of pieces, or a strip, or a double number line. So much for details concerning fractions, otherwise we will digress too far from the 'Vase with Sunflowers'.

Historical Comments

In describing Freudenthal's influence on the development of education for fractions and ratios one must immediately list his monumental 'Didactical Phenomenology of Mathematical Structures' (1983) of which parts were already published as early as 1973. And it is notably the phenomenological reflections about ratios and fractions which have had great influence on the development of learning strands in this area. And yet there was no evidence of direct extraction of education from this analysis. Even stronger, upon first introduction one sooner had the feeling that one was being distracted from it. Take fractions for instance: the enormous variety and complexity of appearances, as described by Freudenthal, had an almost paralysing effect on anyone doing developmental work on this subject. Add to this the emphasis that Freudenthal placed on unicity of individual learning processes – thereto in part inspired by the 'walks with Bastiaan' his grandson – and one can imagine what an enormous burden developmental research in this area was faced with. The result of these efforts is visible in the fundamental work of Streefland (1991a) which in turn adds new elements to the phenomenology, such as fair sharing via ratios and the working with a mediated unit for named fractions. And the notation thereof with letters, plus the (empty) double number line as master model for operating with fractions by Treffers (1991b). In short, Freudenthal indicated where to look and how to look (observation of learning processes) but even he did often not know in advance what would be discovered. His influence was and is fundamental, but also difficult to pinpoint concretely because he was not directly involved in the design of learning courses and did not really have clear cut suggestions for the total education structure for these complex learning courses.

A typical and intriguing remark in this regard was the following: "Why don't we approach fractions in the same way as initial arithmetic, namely depart from elementary context problems and from there try to arrive at formal arithmetic?" A remark that has remained with me until today and a major principle of Wiskobas for the development of ideas about calculating with fractions – leading to ratios, to mediated measures, to named fractions, to explicit notation thereof, to the (double) number line... Therefore indeed comparable to initial arithmetic where also initially one works with named numbers, and the empty number line plays an important role. But it took ten to fifteen years for this encouragement to take concrete shape, one which Freudenthal himself had not anticipated either, but could ultimately appreciate. And still the developers had the feeling that it was what Freudenthal had intended all along. 'Wiskobas and Freudenthal', a complicated relationship – difficult to separate, and certainly in terms of ideas because Freudenthal was a member of the Wiskobas team.

2.3. Measurement and Geometry

'Van Gogh' is mainly about measuring – the measuring of area. Listed below are six aspects of measuring in the Wiskobas program. These apply generally, but we will illustrate them briefly using the example of area.

1. Comprehension of measurement is not automatic. What does it mean to measure area in square meters, kilometres, yards? What is area? The number of plants, for example, says something about the size of a field or else the area. And through this relation to another, more concrete magnitude, area can be made visible, and thus the development of comprehension of measure be promoted.
2. Secondly, 'everyday' measurements must be related to the realm of experience and imagination of the children. What is a square meter, how big is that? Can a non-square also have an area of 1 square meter? How big is a square kilometre, imagine this in your surroundings. How big is 14 square kilometres in that regard? What is the area of the floor of our classroom? Pupils must allow a host of reference points about measure to sink in: how tall are you, how fast is the car going, how many people live in a large city, what is the temperature today... and so very many more. Items like 'Van Gogh' pertain to this. The Wiskobas program places strong emphasis on it. Numbers of measurement can establish arithmetic in the reality of children.
3. Development of strategies of measuring is inherent to measurement. Notably estimation is of crucial importance, especially estimation on the basis of personal reference points. See the trees, the tracks in 'Van Gogh', we already mentioned them in respect to estimation. Estimating and measuring are closely related.
4. The relationships between measures and magnitudes must be examined. Metrics is one of these aspects – see 'Van Gogh'. But also relations between circumference and area, between distance, time and speed, and so forth.
5. Calculating with numbers of measure sets special demands, notably when dealing with rounded off or inaccurate numbers. Compare the indications of 250,000 and 12,820 in the Van Gogh problem, for instance. And the matter of the rounding off of yards and meters – is that allowed? Errors can be blown up when multiplying or dividing (compare 60×70 m – not such a poor estimate but the area is quite far off). Matters to which attention must be devoted in relation to measuring, estimating and calculating.

6. And finally, the collection and processing of numerical data in graphs, tables, diagrams and so forth, something not found in the 'Van Gogh' item. Drawing graphs, interpreting them, all of this is also part of measuring.

All in all measuring offers a natural access to arithmetic, it provides models and context problems which cover a wide area. Various skills and fields are related through measurement – see 'Van Gogh'. The Wiskobas program reflects the great importance of measurement.

There is very little geometry in Van Gogh... Although...? There is an element of spatial orientation in the first problem and certainly in the perspective-problem (De Moor, 1991). The Wiskobas program contains problems concerning projecting and visualizing (1), orienting and localising (2), and shapes and pictures (3). The children are encouraged to construct, calculate and reason.

In the first phase of geometrical instruction the structuring of spatial experiences are mainly visual and implicit: the insights are in fact not expressed in words other than 'I see it this way'. Later intuitive notions must be more considered and better argued by means of language symbols, pictures, reasoning. Then explicit concepts are developed such as angle, perpendicular, direction etc., which possess a certain relation and which can be built up on in secondary education.

In the category of projecting and visualizing, to be somewhat more explicit, problems fit such as:

– Why do shadows become longer when you walk away from a street light and not when you walk away from the sun?
– Why does the moon walk along with you?
– Why does the thumb when held near the eyes change position if you close first one eye and then the other?
– Why do railway tracks converge in the distance?

No easy questions. They especially give an impression of the continuing line of geometry from primary to secondary education. The Wiskivon team (Goddijn, Kindt, Schoemaker and others) produced a great variety of ideas in this field on which Wiskobas also took advantage. Mathematical world orientation requires investigation, explanations of the self-evident, of own experience. There is a great deal of geometry in the Wiskobas program. And that is visible in the publication 'Five years IOWO' (Freudenthal, Janssen and Sweers, 1976) – you can tell at a glance.

Why? Because there is not another part in the program that can be as motivating, that can promote the forming of a mathematical attitude so much, that appeals so strongly to general objectives, as does geometry.

Historical Comments

Freudenthal, the great promoter of geometry education wrote:

> The deductive system of geometry – the ideal for an elite school, was a farce, even for the elite pupils, and for the schools for the masses even less than that. At the end of the sixties I succeeded in formulating what I considered to be the solution: deduction yes, but then local, something the pupils can oversee, not global, as a system thought up by hobby horse riders. (Freudenthal, 1987)

The Wiskobas group and certainly also Wiskivon, the group responsible for secondary education within the IOWO, needed only little urging in this direction: geometry was manifesting itself from reality on every side – geometry as mathematical world orientation which required no extensive didactical phenomenological analysis to develop quickly. The major part of 'Five years IOWO' is about measurement and geometry, for a part integrated with ratios. Freudenthal was also the great promoter behind measuring. He devoted a didactical phenomenological reflection to measuring (1984), to measuring length (1983) and area (1978).

Generally it can be said that the result of Freudenthal's strong fixation on phenomenological diversity of the appearance forms of mathematical structures and on the great individual variety of learning processes was that he was extremely reluctant to dictate learning paths in the steering of education.

In the earlier mentioned evaluation talks about Freudenthal's own estimate of his role in the development of mathematics education in the period 1970–1980, also regarded against the background of recent developments in the eighties, he was very frank in putting forward that his contribution had probably lacked enough of a steering element. Steering both in regard to those giving instruction and the alignment of the instruction material. Seen in that light it is no coincidence that the keyword 'reinvention' from this first major work (1973) is called 'guided reinvention' in his last book (1991).

Summary

Realistic learning strands start with the informal context bound working methods of children, in their personal reality. From there models, schemes, symbolisations are developed which serve as intermediaries to gradually bridge the gap between these start situations and the level of the formal, more general subject related operations. The learning of the tables for instance starts with the item about the number of days in so many weeks. Visualising via the time line and the calendar leads directly to the (empty one sided and double) number line and the rectangle. Two master models which are of essential importance for insight in multiplication, also in the areas of ratios, fractions and measuring. And for learning long division a dynamic context problem about the transportation of passengers with buses serves as model context or paradigm that provides a concrete basis for formal operation. The same occurs for fractions with problems about fair sharing and measuring, whereby the language of fractions is developed and where comparing and operating are given a concrete basis of orientation.

With this specific vertical mathematising combination of the starting situation from personal reality and the subsequent model forming, schematising and symbolising, and added to the horizontal theme-approach, all of this together distinguishes the realistic approach from the mechanistic, structuralistic and empiristic domain theories and practices. (See Treffers, 1987, 1991b). Freudenthal laid the foundation for this didactical realism and determined the development of concrete education for various main learning strands, but more indirectly than directly because he himself did not design or outline themes and learning strands.

3. INTEGRAL EDUCATION DEVELOPMENT

3.1. Historical Comments

As mentioned before, Wiskobas already existed before the founding of the IOWO in 1971. And the idea of integral education development already existed before Wiskobas became a department of the IOWO. Wiskobas pioneers Wijdeveld and Goffree more or less brought it along when they joined the IOWO, or really, acted as co-founders. The idea to approach innovation of mathematics education on a wide front – via training institutions of future teachers, via in-service training, training of counselors and instructors, research, development, via textbooks, and curriculum development – this was an idea that Freudenthal was most keen on. The more so when he established in 1970 that the Wiskobas movement was not only viable, but that is was also well organized in one national and several regional working groups. When the IOWO was founded he wholeheartedly gave his best efforts towards coordinating and realizing the integral approach of education development in the best possible manner. And it is predominantly due to Freudenthal's authority, status and influence that this integral approach did indeed bear fruits. Even though his critical attitude with regard to educational research, curriculum and test development did guarantee that the discussions were always lively – to put it mildly. The consequence of the broad approach was in any case that at the end of the eighties the development of textbooks, tests and curricula were well attuned to the national final standards and to the (informal) national curriculum (de Jong, 1986). While in many other countries (Ernest, 1991) any such attunement is lacking and the textbooks, the national curriculum and the tests explicitly lack the concrete views of progressive educators. For a description of this process of integral education development and the construction of an adequate infrastructure it would lead too far a field to go into this very complex process at this time. Freudenthal does dwell on the matter rather extensively in his last book (1991) in his outline of 'The Landscape of Mathematics Education', in which he concentrates on theory forming, research and dissemination. These three aspects become visible most explicitly in developmental research. And what Freudenthal writes about it is so important that I would briefly call attention to it here.

3.2. Developmental Research

Developmental research denotes a comprehensive program that is directed at innovation and improvement of education. It is research with an important development component: it is not merely established how things 'are' in existing education (especially research on errors is very popular in this respect) but much more how things should be, and one develops education that suits these findings (in a theoretical and practical sense). Development and research take place in an integrated cyclistic process. And the developer must report on that process: the own learning process therein, and the rational and empiric justification of the developed product must be made clear, so that others can also benefit from it in their development, research and teaching (also see Gravemeijer, 1989). Developmental research comprises much more therefore than empiric testing, but rather all sorts of aspects of innovation, among them that of dissemination – this as opposed to the more common research. Freudenthal observed that developmental research of this kind does not fit well in the current RD and D model, and therefore also not in regular educational research.

4. CONCLUSION

In July 1990 Freudenthal pointed 'his' associates of the OW & OC (currently the Freudenthal Institute) to the following article in a national newspaper:

> THE POLISH KNOWN AS HARD WORKERS
>
> Every year tens of thousands of Poles come to The Netherlands to work in the flower bulb industry for a few months. (...) This is the fourth time for Zygmunt. He has worked both in the fields and in greenhouses. Currently he is working in the transportation department of a company at the flower auction. "I load up the trucks, that is heavy work. On average I work 220 hours per week. That's no problem, because that's the way to make money" says Zygmunt (NRC/Handelsblad, July 1990)

$>$ The question is: Is that possible, to work 220 hours per week?

Children of 9 and 10 (start of grade 4) appear to answer this question in many different ways:

- "I have no idea, I guessed 'yes'."
- "No, because my mother already works 180 hours per week. If he worked even more hours than he already does she would be busy all day."

Interesting answers because it proves that children do not automatically mathematise the problem, for example by coming up with the given that a day has 24 hours and a week 7 days. Children of 5 and 6 do not yet have this knowledge. And apparently neither do some nine year olds. In any case they do not already use reference numbers of this kind.

Even if they do have that knowledge, the reasonings and calculations differ substantially. Take the reasoning which departs from an 'exhausting' work week, therefore of '7×24' to refute the result of 220 hours. The calculations of '7×24' vary greatly: repeated counting $24 + 24 + 24 + 24 + 24 + 24 + 24$ by various calculation methods; front-end method of stylised mental arithmetic $7 \times 24 = 7 \times 20 + 7 \times 4$, and the end-front method of the standard algorithm.

Characteristic for realistic mathematics education is that it puts such diversity to good use. For instance the front-end method can be presented as an abbreviation of repeated addition and also as basis for column arithmetic.

This does not exhaust the source of instruction possibilities: efficient arithmetic and estimation can also be used as examples:

- A day has 24 hours, so I take $7 \times 25 = 175$; minus 7 hours = 168 hours, the answer is 'no' (Figure 4).
- No, because 10 workdays is already 240 hours, so 220 hours would be more than 9 (Figure 5).

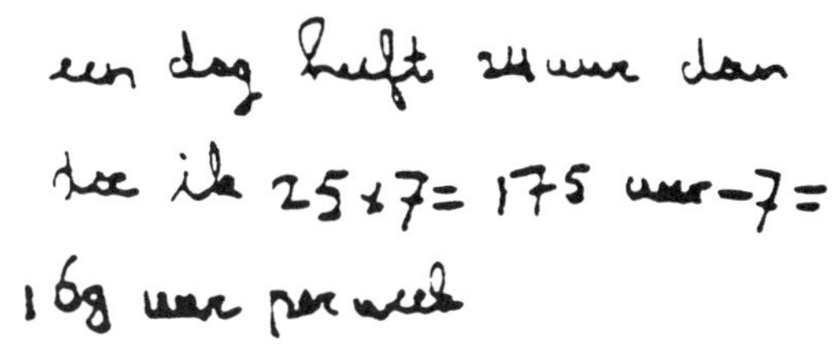

Fig. 4.

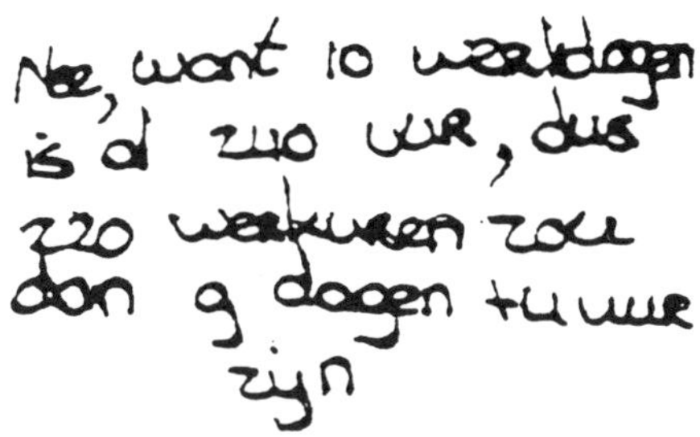

Fig. 5.

– No, because $220 : 7 = 31\frac{2}{7}$, but a day only has 24 hours (Figure 6).

The example shows how many learning possibilities there are in that one context problem. Not that all of these opportunities must indeed be employed, but one or several should be used, such as the discussion and analysing of the relations between the various solution methods of (stylised) mental arithmetic, estimation and column arithmetic.

Mathematics can be developed from (personal) reality in natural manner: the formal rules and procedures can be derived from the informal working methods of the pupils – that is the heart of didactic realism.

In the first part of 'Revisiting Mathematics Education' (1991) Freudenthal gives an in-depth analysis of the specific 'common sense' character of mathematics. 'Specific' because according to Freudenthal the mathematical activity distinguishes itself there from natural sciences. Elementary mathematics was invented many times over in the world in an independent manner, but there was only one Newton, Mathematics develops according to a gradual ongoing process. But primitive mathematical notions retain their original value as knowledge, skill and insight increase, while the natural science intuitions often prove not to hold after a time. In short, mathematics is excellently suited for gradual rediscovery, accord-

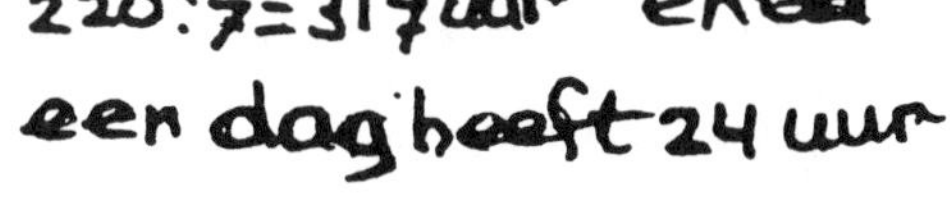

Fig. 6.

ing to Freudenthal (also see Goffree, 1992). In the second part of his last book the didactical principles of 'guided reinvention' are described, and in the third part Freudenthal concentrates on theory forming, research and dissemination. A few elements from these two parts about didactics and innovation are brought forward in this article about realistic mathematics education:

> to bear witness for the product to make it plausible and transferable, which demands an attitude of self-examination on the part of the developmental researcher: a state of permanent reflection. (Freudenthal, 1991).

Much about Freudenthal and his relationship with Wiskobas has not been highlighted: his ideas about levels in the learning process, his views on the issue of assessment, didactical classification methods, the educational-theoretical framework of realistic mathematics education and so on. But most of all his personableness, his intriguing off-hand remarks, his elaborations on seemingly futile observations of learning processes, his passion, his ability to use simple examples to show that mathematics is everywhere, his attention for both 'simple' and 'complex' forms of mathematics, his respect for practitioners of teaching, his admiration for a point well argued, his style of writing, his undisguised disapproval of developers and researchers who do not do justice to mathematics nor to the child, do no justice to mathematics as a human activity, to 'mathematics as an educational task' ...

One of the children answered the Zygmunt-item as follows: "Yes, 220 hours per week is possible, because you can load trucks and let the tulip fields grow at the same time".

Indeed, that is possible. Because that is also how Freudenthal worked... And at the same time he let grow the tulip fields of realistic mathematics education. It was a privilege working together with him in the Wiskobas team.

NOTES

[1] I would like to thank Aad Goddijn, Fred Goffree, Rob de Jong, Edu Wijdeveld and other colleagues for their useful comments on the first draft of this article.

[2] Wiskobas is an acronym for the Dutch translation of 'mathematics in primary school. The Wiskobas project started in 1968 and was later adopted by the IOWO (Institute for the Development of Mathematics Education). When this article mentions Wiskobas it refers to the project team that operated within the IOWO and after its termination continued its work at other institutes, and in the course of time was even expanded somewhat. When the text refers to the Wiskobas program, this refers to the informal National Curriculum for arithmetic in primary school that was compiled at the end of the eighties at the initiative of Wiskobas staff, by a wide group of progressive educators to whom, for this occasion, we will refer as the Wiskobas movement. Wiskobas spearheaded this movement. Freudenthal was the most prominent member of this innovative group.

[3] The department for secondary education within the IOWO was called Wiskivon. It originated somewhat later than Wiskobas. Freudenthal's direct involvement with this group was initially less intensive than it was with Wiskobas. But his ideas also found their way into secondary education via Wiskivon, both through the SLO (Curriculum Development Foundation) and the research group OW & OC (currently the Freudenthal Institute).

[4] This conversation was held on 11 August 1989, between H. Freudenthal and the author.

REFERENCES

Brink, J. van den: 1991, 'Realistic Arithmetic Education for Young children', in Streefland, L. (ed.), *Realistic Mathematics Education in Primary School*, Freudenthal Institute/CD-β, Utrecht, pp. 77–92.
Ernest, P.: 1991, *The Philosophy of Mathematics Education*, Palmer, London.
Freudenthal, H.: 1973, *Mathematics as an Educational Task*, D. Reidel Publ. Co., Dordrecht.
Freudenthal, H.: 1977, 'Wiskunde Onderwijs anno 2000 – Afscheidsrede IOWO', *Euclides* **52**, 290–295.
Freudenthal, H.: 1978, 'Oppervlakte als verschijnsel benaderd', in R. de Jong (ed.), *Oppervlakte*, IOWO, Utrecht, pp. 109–120.
Freudenthal, H.: 1983, *Didactical Phenomenology of Mathematical Structures*, D. Reidel Publ. Co., Dordrecht.
Freudenthal, H.: 1987, *Schrijf dat op Hans*, Meulenhoff, Amsterdam.
Freudenthal, H.: 1991, *Revisiting Mathematics Education. China Lectures*, Kluwer Acad. Publ., Dordrecht.
Freudenthal, H., Janssen G.M. and Sweers, WE.J. (eds.): '1976, IOWO-snapshots', *Educational Studies in Mathematics*, **7** (3), 188–367.
Goffree, F.: 1992, *HF: Working on Mathematics Education. (paper).*
Gravemeijer, K., van den Heuvel, M. and Streefland, L.: 1990, *Contexts, Free productions, Tests and Geometry in Realistic Mathematics Education*, OW & OC, Utrecht.
Gravemeijer, K.: 1989. *Evaluatieonderzoek vanuit ontwikkelingsperspectief* (paper).
Heege, H. ter: 1985, 'The acquisition of basic multiplication skills', *Educational Studies in Mathematics* **16**, 375–389.
Jong, R. de: 1986, *Wiskobas in Methoden*, OW & OC, Utrecht.
Lange, J. de: 1979, 'Contextuele problemen', *Euclides* **55**, 50–60.
Lange, J. de: 1987, *Mathematics, Insight and Meaning*, OW & OC, Utrecht.
Moor, E. de and Streefland, L.: 1991, 'Van Gogh in het graan', *Tijdschrift voor Nascholing en Onderzoek van het Reken- Wiskundeonderwijs*, **10** (1), 61–64.
Moor, E. de: 'Geometry-instruction (age 4–14) in the Netherlands – the realistic approach; in L. Streefland (ed.), *Realistic Mathematics Education in Primary School*, Freudenthal Institute, Utrecht, pp. 119–139.
Streefland, L.: 1991a, *Fractions in realistic Mathematics Education*, Kluwer Acad. Publ., Dordrecht.
Streefland, L.: 1991b, 'Fractions an integrated approach', in Streefland, L. (ed.), *Realistic Mathematics Education in Primary School*, Freudenthal Institute/CD-β, Utrecht, pp. 93–118.
Treffers, A.: 1987, *Three dimensions. A Model of Goal and Theory Description in Mathematics Instruction – The Wiskobas Project*, D. Reidel Publ. Co., Dordrecht.
Treffers, A.: 1991a, 'Meeting Innumeracy at Primary School', *Educational Studies in Mathematics*, **22**, 333–352.
Treffers, A.: 1991b, 'Didactical Background of a Mathematics Program for Primary Education', in Streefland, L. (ed.), *Realistic Mathematics Education in Primary School*, Freudenthal Institute/CD-β, Utrecht, pp. 21–56.
Treffers, A., De Moor E.: 1990, *Proeve van een Nationaal Programma voor het Reken-wiskundeonderwijs op de Basisschool*, (deel II), Zwijsen, Tilburg.
Whitney, H.: 1988, *Mathematical Reasoning, Early Grades*, Princeton (paper).

Freudenthal Institute,
Utrecht University,
The Netherlands

(H.F. in: Editorial letter 5 July 1978)

In the case at hand however, these so-called heuristics are dependent in a more awkward manner. I mean, dependent on the specific problem to be solved, and therefore I call them clues – a clue for every problem and a problem for every clue. It reminds me of teaching techniques such as: '3+2 are ...?' No answer. "Think". No answer. "F...". "Four". "No". 'Five". "Very good". Of course this is an exaggeration but you will understand what I mean. I am not at all surprised that the experimental group is superior to the control groups.

Then for the subject of 'integration'. Of course, if you train students in a subject like this you should do it methodically. Unfortunately the first question is whether students should be trained in a subject like this, or whether there are not more important subjects in mathematics to be taught, regardless the career the student has in mind. Do not forget that the whole teaching subject of calculating indefinite integrals was invented to be tested out. I wonder whether it still exists at most universities.

As a matter of fact almost all efforts to develop diagnostic tools in teaching arithmetic show the same features as the project you have sent me. One of the most outstanding and worst among them is the Swedish project. Any physician who would make such a superficial diagnosis, from mere symptoms, would rightly be labelled a quack. I have never seen any benefit in ascertaining whether a child is not able to perform this or that partial arithmetical operation even though all of them might have been carefully selected. It is not important to know which problems a child cannot solve if no effort is made to ascertain why it cannot. In order to know why children fail, one must observe them. In medicine it is called aetiology if one tries to find the causes for failure. I am sure this is the right way to make a diagnosis because in the course of the years I have had some experience with remedial work, and during the last few years I have cooperated with a group of psychologists, who in fact have singled out a few important causes of failure, which in a sense are still symptoms, though more profound than those one discovers through armchair analysis of arithmetical problems rather than by observing the learning processes. The tests proposed in your project do not even apply to low achievement computers, let alone to human low achievers. I suppose that serious work on diagnosing low achievement in arithmetics has been done and is still being done at quite a few places in the world, but unfortunately it is lost in the flood of irresponsible work. I call it irresponsible because it seduces good teachers, who try to understand why certain children fail, to believe that superficial testing may supersede careful observations, and because it teaches average teachers and student teachers to suffice by looking for mere symptoms instead of looking for more profound reasons for failure. In the research I alluded to, it appears that more than 50% of serious cases of low achievement are due to the lack of any decent instruction, which for that matter is the consequence of poor teacher training. Those poor teachers, rather than learning to look for possible causes of failure, were educated to believe in blind tests.

L. STREEFLAND

THE DESIGN OF A MATHEMATICS COURSE
A THEORETICAL REFLECTION

None of our theories remain valid for unlimited time, the images of our reflection will change, concepts which were elevated to general principle on the basis of their success will have to be abandoned, but we are left with the certainty of being able to order and predict large groups of phenomena (H.B.G. Casimir, Waarheid en visie, p. 34).

ABSTRACT. Realistic Mathematics Education has its conceptual framework, that is a theory. This theory evolved from development and research in mathematics education.

A course on fractions and many other examples will be analysed in depth with the realistic theory in the background. The leading question will be: "What strategies did the designers of the exemplary educational material use?" Revealing those strategies and making the user conscious of them enables the user to apply them henceforth as tactics for the design in advance.

In consequence an important theoretical change of perspective looms ahead. Where the theory was first an *after*-image, it can now act as a *pre*-image, i.e. as a *model for* realistic mathematics education in advance. This is because the tools to farm the theoretical field have been revealed. The question of *how* to apply the theory has been answered. Freudenthal's ideas provided for a considerable part of the means to make this theoretical step forward.

INTRODUCTION AND OVERVIEW

What kind of mathematical-didactical attitude befits the undertaking in the title? What does such an attitude entail? What is didactics? How does it originate? What is a learning course? How does one design a learning course? Is this issue not overly ambitious? Is a learning course ever finished? Note that the posing of a question does not necessary also mean giving the answer. New questions arise from the very asking as it were. They form the leitmotif for this article. The context is the activity of didactising, although this can only be denoted in a preliminary sense.

It is the activity of them that teach mathematics, design it, do developmental research on the subject and compile suitable assessments and construct theory for it. It is these activities that bring forth didactics. Therefore, didactics is about the arrangement of the contents of mathematics instruction, broadly speaking. But it is also the science of structuring. For example, how one can set out fractions for 10 to 14 year- olds thus that the best imaginable learning results can be achieved over the full expanse of a heterogeneous group, the right tests to assess the same included. Let us therefore didactisise! Fortunately we need

Educational Studies in Mathematics **25:** 109–135, 1993.

not start off empty handed. The posthumous publication of the book 'Revisiting Mathematics Education' by Freudenthal (1991) is about didactising. Freudenthal links didactising to mathematizing. He sees a clear parallel between the two activities, if not a striking resemblance. This resemblance had already been observed earlier (Freudenthal, 1981; Streefland, 1985, Treffers, 1987).

In the last two decades a significant number of didactical tools have been developed for the designer to use. Freudenthal's didactical phenomenology for instance (1983), together with the other building blocks for realistic mathematics instruction. And of course especially the material produced by Wiskobas, Wiskunde 12–16, HAWEX and HEWET. And all of this with in its wake the text books that have appeared on the market since.

There is also the substantial didactical repertory for the educational designer that Goffree and Stroomberg (1991) discuss. They reflect on – what they call – the architecture of lessons and learning courses. They take a look at a number of major design strategies, among them Freudenthal's phenomenological approach.

Is there any sense then in prolonging this article? It would seem that is has all been said and done. Yet I would answer the question in the affirmative, without any reservation, because mathematics instruction as a science is a long way from being completed, nor can it be! This despite the many publications by Freudenthal and others who have supplied the building blocks for such a science. The creation of this discipline has in fact only just begun.

The first part of this article is about the level of the theoretical framework for realistic mathematics instruction. Globally speaking this concerns mathematics education searching for its problems and questions in reality and its phenomena in order to serve both as a source for mathematics and as a field of application. Next, a learning course on fractions is subjected to a closer analysis. Through such analysis the strategies of the designer come to light. The question is whether these strategies are bound to the context of the course involved or not. In order to answer this question examples are taken from other courses, also to eventually discover different strategies of design. 'Mathematics course' from the title thus acquires a more general meaning. This awareness allows these strategies to crystallize out as the method of operation of the designer. They become a part of his tactics. This will also be shown by examples.

Subsequently the theory for realistic mathematics instruction can be brought to a higher level. Furthermore, theory has been made operational because the tools that are needed to treat the current theoretical field have been discovered.

MENTAL OBJECTS, CONCEPTS AND THEORY

Concept of Fractions

Suppose a pupil does not yet have a full conception of fraction that is in line with the systematics of the subject, but that he or she is no longer totally undetached from it. Then this pupil is 'somewhere' on the way towards understanding the concept. In the terms of Freudenthal (1983) the pupil is working on the constitution of the mental object of fraction. I am looking at 'fraction' somewhat closer to instruction now. What is happening in that regard? A great deal, these days. Take distribution situations such as '4 children divide up 3 chocolate bars among themselves, how much will each get?' The careful bar-after-bar, the more daring 'first two bars and then one', and variations thereof, produce a rich source of varied material:

$\frac{1}{4}b + \frac{1}{4}b + \frac{1}{4}b$, $3 \times \frac{1}{4}b$, $\frac{3}{4}b$, $\frac{1}{2}b + \frac{1}{4}b$, ... (Streefland, 1991a). The describer '$\frac{3}{4}b$', a part-whole relationship generated by division, originates in different ways. '$\frac{3}{4}b$' is already outlined in '3 bars and 4 children'. Afterwards this relationship is established and described. At first the only objective is dividing and distributing. Nothing more. For the child that is learning there does not yet have to be any notion about fractions. That follows later on when such describers and their mutual relationships themselves become the subject of reflection. The part-whole relationship as such, the first aspect of the concept of fraction, is therefore an a posteriori notion.

Realistic Theory

Comparable to the foregoing is how the realistic theory of Wiskobas came into being, naturally not to be compared in terms of complexity and time, but strikingly similar in terms of the nature of the process. Take the role of history. In ancient Egypt computation with fractions originated from dividing (and measuring) fairly, as evidenced from the Papyrus Rhind (1700 B.C.). Even today this document can be a rich source of inspiration for education, for anyone with an eye for it.

Much the same applies for realistic theory, although its history is of much more recent date. Realism can be found in the bountiful tradition of German arithmetic didactics, as one find in the work of Kühnel (1916) for instance. Freudenthal (1968; p. 7) anticipates on this where the developments in The Netherlands are concerned:

> In its first principles mathematics means mathematizing reality, and for most of its users this is also the final aspect of mathematics. For a few this extends to mathematizing mathematics itself.

'Mathematizing reality' and 'mathematizing mathematics'. Is Freudenthal already referring here to what Treffers (1987) would later call 'horizontal' and 'vertical mathematizing', with this nuance that the mathematizing of mathematics itself, regarded from the pupil, does not at all have to be the privilege of a chosen few? But, I digress. Or perhaps not. At issue is a theoretical framework as the fruits of a learning process, just as growing insight in fraction comes about in a learning process. And is it not so in both cases that the person who is learning must (sometimes) come to such insight through the eyes of others? Whatever the case, in the Wiskobas material, W 12–16, etc, realistic theory was there, submerged, and it proved possible to bring it to the surface:

- concretizing to construct (utilizing reality as a source and application area)
- interweaving learning strands to structure, etc. (Especially the interweaving of learning strands will in this connection demand much attention).

An important precondition for such a learning process has been that the perspective on mathematics as a human activity lay at the basis of the development of the material and the mental constitution of the theoretical framework.

The theoretical framework itself was also, just as the fraction in the part-whole relationship, a posteriori determined; therefore of a descriptive nature (Treffers, 1987; p. 242, Gravemeijer, 1987).

Operational Model?

One could take the process of the forming of fraction as mental object as a model for the forming of theory in realistic math instruction, if this is not being too presumptuous, that is. Because, 'fraction' encompasses much more than a 'part-whole' relationship. This notion afterwards must be active before-hand in order to be able to apply it in related or new situations. Take something such as 'what does $\frac{3}{4}$ part of an 80 cent chocolate bar cost? '$\frac{3}{4}$ part' will first have to be applied on the price of 80 cents to determine the price of $\frac{3}{4}$ part. There is a shift from part-whole to operator. In general a transition such as this goes from:

...examining a given thing and stating its properties... to
...making a thing with prescribed properties... (Freudenthal, 1991, p. 94).

And that is what happens in the example, the relationship three/quarters must as yet be made in the price of 80 cents. This involves the raising of level. The operational remains in the fore also for the part-whole relationship, because it is based on fraction-generating fair sharing. At issue here however is operating at still a higher level, one that builds on fair sharing with its generated relationships.

An operational concept of fraction involves comparison, ordering, determining difference, adding together etc; operations which ultimately – possibly – can be executed according to the rules and laws of formal subject systematics.

That – in a nutshell – is how the learning process for fraction progresses, although the outline is extremely one-sided. In essence it concerns the transition from describer afterwards to operator beforehand, from descriptive notion to rule or standard. And it is not enough to establish that this is how it should be. One will also have to indicate how pupils can come this far, complete with heuristics and rules that are made concrete in a sample learning course.

Should the same then not also apply for realistic theory? It does not suffice to say that it comes down to concretizing in order to construct, model forms to abstract, etc. It will also have to be indicated how this can be achieved. In other words: theory must (as yet) be made operational.

Theory Level and Level Theory

Is the theoretical framework for realistic mathematics instruction not operational?, one might ask. Has 'concretizing' not been made applicable through specifications such as 'departing from contexts' or 'using reality as a source and application area'?

Basic characteristics such as these, although concretized, are of a too non-committal level to be able to answer the how-question. This despite the fact that there is practice of it which gives a certain normative substance to the basic characteristics.

Moreover, this general level (still) leaves too much open. Is it not true, speaking in second hand terms, that reality as a source and application area, that frequent application of schemes, models, etc. and the interweaving of learning strands point to the learning material? And the same applies for own productions, albeit with the pretension of reflection as an effect of learning. The interaction in conclusion refers to teaching itself. And it is precisely on this point that realistic theory has hardly been specified. One should bear in mind in this regard that this pillar of reflection rests on the notions of horizontal and vertical mathematizing. Van Hiele's level theory structures the mathematizing process to levels to be progressed. (Treffers, 1987).

But,...has it thus been indicated how reality can or even must be utilized, how one must deal with contexts, which models can be developed, where levels can be raised in the learning process and how one might achieve this, how and where one would interweave learning courses, where one would place production items and how these in turn would merge in the total process of the learning course that is emerging, what the position of practising is in problem-oriented instruction and how this can for example be interwoven with the learning strand that is aimed at growing insight, ...?

But, ... has it thus been indicated how...? The answer is an unequivocal: no! For the simple reason that the theoretical framework for realistic mathematics instruction is not (yet) operational. It is still only illustratively operational, due to the material that has been developed and tested. This is pregnant with theory but only still implicit as far as the higher levels are concerned. Regarded in terms of theoretical level, realistic theory still finds itself at the concrete level, or in other words, the forming of theory is still at the stage of the constitution of mental objects. If we truly want to answer the question posed above then we will have to free ourselves from this concrete level. It must then be brought to light how to operate on the concrete building blocks for realistic mathematics education theory and on their mutual relations. In the terms of Freudenthal's posthumous book this comes down to the vertical application of the realistic, theoretical framework that originated from horizontal didactising. In other words, it will have to be paradigmatised, i.e. be made from a post-image to a pre-image. In a metatheoretical sense levels will therefore have to emerge which will supersede that which is domain- specific. (Whereby, among other, connections with other disciplines become visible. A point which will not be pursued further in this current presentation). To say that:

> ...pupils must constantly have the opportunity...to reflect on learning strands...and to anticipate...
> ...mathematics education should by nature be interactive...
> ...learning strands must, where possible, be interwoven with each other...(Treffers, 1991, pp. 25 and 26),

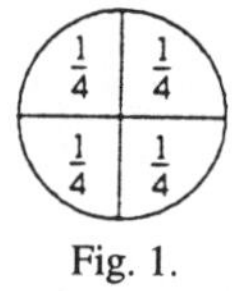

Fig. 1.

is not enough, certainly not when the 'how' is only indicated in the global terms of the considered theoretical framework. To give an example, that own productions and conflicts can serve for reflection.

What I have in mind reaches further and delves deeper.

Anyone who switches from a posteriori conclusion to a priori prescription will have to indicate more precisely how the conditions to achieve this can be created in the material (or in the interaction!).

The operationalisation of realistic theory will be the most convincing if it is made applicable as design repertory. However, for a long time Freudenthal was reticent about theory forming for mathematics instruction. According to him the danger of rashness lurked. Later on he became somewhat more lenient:

> From the desire to understand practice grows theory that in turn perfects practice and improves it, and – if described efficiently enough to be used again – also influences the practice of outsiders who have not themselves experienced the development process directly. That is the purpose and objective of theory. (1987, p. 14).

Watch out! Freudenthal says in his leniency: As long as it starts with the theory of...! This quote also reflects the purpose and objective of this article. Let me therefore turn to the practice of designing and do it thus that the results thereof can again be applied by others, by designers for instance.

DESIGN OF A LEARNING COURSE, FRACTIONS AS PARADIGM

Traditionally, fraction-generating fair sharing occurs within the unit (Figure 1). The parts are named and calculation with fractions can commence.

$\frac{1}{4} + \frac{1}{4} = 2 \times \frac{1}{4} = \frac{2}{4} = \frac{1}{2}$ etc. This approach has proved not to work for a great many children, if not for all. In more recent approaches there has been a switch among other to distribution situations in following of the example from the Papyrus Rhind, e.g. 'divide 3 chocolate bars among 4 children, how much will each child get?'. Insofar as these were seen the didactical consequences this had were considerable. I will touch on them step-wise:

- a numberless or qualitative approach makes possible an unhindered orientation on suchlike situations: some children divide some bars, how much will each one get, more or less than a whole bar, what will happen if a few others join the group of sharers, will they get more or less then?;
- the comparison of situations and estimation become meaningful; e.g. for '3 bars, 4 children', 'does everyone get more or less than $\frac{1}{2}$ of a bar or than a whole?;

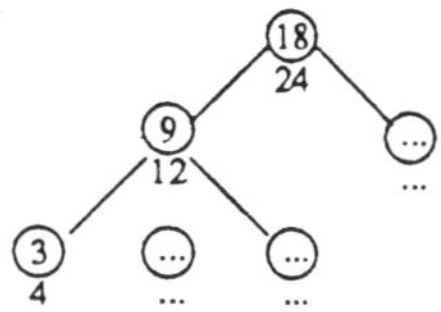

Fig. 2.

- because there are several units and because the dividers are explicitly participants, the division can be done in various manners, for instance first two bars with everyone getting half a bar and then the third bar with everyone getting a quarter bar, or bar by bar, with $\frac{1}{2}$ bar each, repeated three times;
- the result is that the first notions of the concept of fraction go hand in hand with the (informal) operation and furthermore that the phenomenon of equivalence, as in the traditional approach, no longer needs to be exacted from the material, but comes naturally from the division; portions like $\frac{1}{2}b+\frac{1}{4}b$ and $\frac{1}{4}b + \frac{1}{4}b + \frac{1}{4}b$ are the same, therefore also $\frac{1}{2}b$ and $\frac{1}{4}b + \frac{1}{4}b$, in short: the equivalence becomes more widely operational in consequence;
- the different manners of division can occur spontaneously in a year group of pupils, but one can also steer this process by letting it take place in a coffee shop or some such place; the speed of the kitchen and the service in consequence – first serve two and later one, or one at a time – then determine how things are divided at the table;
- by the presentation of the dividers in the situation attention can also – albeit with the necessary limitations – be drawn to how they are seated at the table; a separate symbol can be thought up for this purpose (by the pupils). For example $\underset{8}{\textcircled{6}}$, which stands for 6 bars or something else on the table and 8 children seated around, which leads to $\underset{4}{\textcircled{3}}$ $\underset{4}{\textcircled{3}}$, two tables, each with 4 children and therefore 3 bars on each table $\underset{4}{\textcircled{3}}$ $\underset{4}{\textcircled{3}}$, in order to divide fairly. In this manner the fractions are closely related to ratios; moreover, the seating arrangement can take place in a progressing process of schematisation by applying all sorts of abbreviation to the scheme. Equivalent situations can be brought forward (Figure 2). Via the various forms of dividing, each person's portion per table can be determined in the scheme. A close link will grow between situation $\underset{4}{\textcircled{3}}$ and the portion $(\frac{3}{4}b)$.

Following the line for fractions and leaving the connection with ratios – insofar as possible – for what it is for the moment, the following building blocks are of some import.

- departing from someone's portion, e.g. $\frac{1}{2}b$, one can look back at the situation of origin, $\underset{4}{\textcircled{1}}$, $\underset{8}{\textcircled{2}}$, $\underset{12}{\textcircled{3}}$, ...; also for constituent portions this can be done, with the conflict problem such as: can someone with $\frac{1}{2}b + \frac{1}{3}b$ have been sitting at table $\underset{5}{\textcircled{2}}$?;
- varied distribution results in a division situation with a number of descriptive relationships, $\frac{1}{2}b + \frac{1}{4}b$, $\frac{1}{4}b + \frac{1}{4}b$, $3 \times \frac{1}{4}b$ etc., all standing for each person's

portion of $\frac{3}{4}b$. This entity of descriptions for one and the same situation was given the name mini- monograph (Streefland, 1991a); referring back to division situations the pupils can (re)construct mini-monographs, whereby in first instance the division in the situation and later on all kinds of other considerations lead to the production of relationships, e.g. by applying commutativity, '$\frac{1}{2}b + \frac{1}{4}b$ is replaced by '$\frac{1}{4}b + \frac{1}{2}b$' or by considerations of equality, e.g. adding $3 \times \frac{1}{4}b$ to $3 \times \frac{1}{2}b$'.

- in context situations in which division (or measuring) takes place, the objects can be represented by their shapes and by their values, (length, weight, price, etc); the forms are stylized into visual models, to which the values can then be linked and moreover be maintained in the notation.

$\frac{1}{4}b$	$\frac{1}{2}b$	$(\frac{3}{4}b)$	$1b$
(20c)	(40c)	(60c)	(80c)

- $\frac{1}{2}b + \frac{1}{4}b$ with a price per bar of 80 cents then becomes $\frac{1}{2}(80c) + \frac{1}{4}(80c)$. By way of the result of 60c one arrives at $\frac{3}{4}b$. In this manner these values (of magnitude) become excellent intermediaries for the execution of the operations with fractions via a roundabout manner.

I will restrict myself to these building blocks for a learning course on fractions.

REFLECTION

What can be learned from the above for the benefit of designing learning courses in general? Quite a bit. Let me take the different building blocks one at a time.

The qualitative approach in the comparison of situations and in estimation, in anticipation of the division and each person's portion, are essential. The pupils are given the opportunity to structure preliminarily without immediately being tasked with the total complexity this involves. As far as estimation is concerned, an awareness for workable reference points is developed from the outset (in this case $\frac{1}{2}b$).

Now, designing is the execution of a thought experiment of teaching and learning both! Approaching an are a means that a designer, the eye directed on the learner, is alert to

– catching the right degree of precision.

Because 'qualitative approach' and 'estimation' have gradually become common property, the example as a result of design is not yet very convincing. Therefore I will here – and in several other instances in other connections and also from other learning courses – add an example of greater authority.

The Shoe Soles of the Giant

Giant Egbert, the main character in 'with the giant's regards' (De Jong (ed.), 1978), is four times the (linear) size of the teacher. At a certain moment there is the problem which deals with the size of the soles of Egbert's shoes and of

Fig. 3.

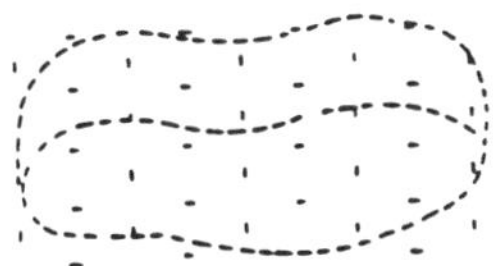

Fig. 4.

his handkerchief (or newspaper). Should the designer now include in his (draft) design the sole or the handkerchief, or first the sole and then the handkerchief, or the other way around? That is the question at issue here. At the time one opted for 'the sole first', comparing it to that of the teacher, estimating 'so many soles of the teacher', construction drawings, etc. (Figures 3 and 4), and later the handkerchief.

Objections were raised against this course of affairs by an international forum of didacticians (Streefland, 1984a). At this level (third grade) the children are not yet capable of 'proving' with the shape of the shoe sole that 4 (in the length) × 4 (in the width) = 16 soles of the teacher's, so it was argued. In this manner it remains a supposition and the mathematics cannot be rounded off. It would have been better to choose for the handkerchief (or newspaper) because it can be 'proved' that that would require 16 handkerchiefs (or newspapers).

The learning process showed however that the designer (Hans ter Heege) had got it right. It appeared that it was not self evident for the children that linear enlargement works in two directions where this concerns flat shapes and that therefore the giant sole had to result in 4 by 4 human shoe soles. Through the activity of estimation and the anticipation on the ultimate construction, whereby two soles in the width (Figure 3) were in first instance considered sufficient, the mentioned insight emerged, via the conflict with the too narrow design: linear enlargement works two ways – in this case. With a handkerchief or newspaper the shape exacts this two-dimensional action so specifically that the necessary attention and consciousness will (possibly) not be raised (The shapes are too convincing in this case).

Hence, in this activity the designer caught the right degree of precision to allow the intended insight to take hold. Equipped with this insight the rest of the learning route for linear enlargement could be anticipated with confidence.

After this example I wish to return to the fraction line.

Exploiting the Context

When departing from divide (and measurement) situations, concept and treatment go hand in hand. The gap between the two, characteristic for mechanistic courses, is thereby closed. The orientation basis which is laid in this manner for fractions,

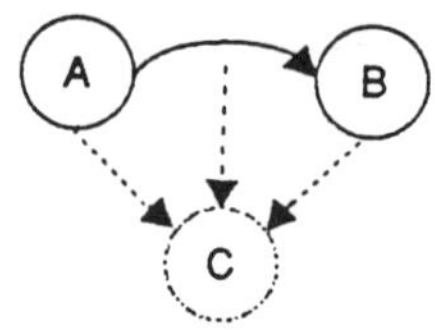

Fig. 5. Directed shifting (ds).

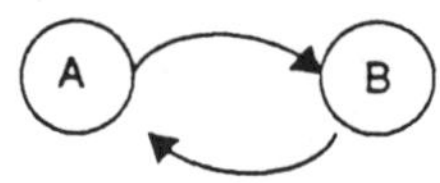

Fig. 6. Reciprocal shifting (rs).

has a strong anticipatory action, both in terms of the key property of 'equivalence' as well as (informal) operation itself. To allow this anticipatory orientation to be as complete as possible, the designer on a number of occasions shifts the attention within the context. That is the case for instance in the transition from the table situation (in the coffee shop) to the kitchen (serving) to be able to steer fair dividing in different ways, thus to be able to elicit as rich a variation of material for fractions as the situation allows.

Shifting can take place in different ways, namely,

(a) 'directed', which means from 'the one' (A)/(phenomenon, point of view, feature, ...) to 'the other' (B), while keeping an eye on or considering C (Figure 5).

(b) 'reciprocal', that is from 'the one' (A) to 'the other' (B), in order to go back to A (Figure 6).

(c) 'parallel', that is from 'the one' (A) to 'the other' (B), while trasfering 'something' from A to B. (Figure 7).

(based on Freudenthal, 1991, p. 105.)

A final, subtle form is

(d) 'reflexive shifting', that is from 'the one' (A) to itself, while considering a particular aim. (Figure 8).

This form, which is a special case of directed shifting will not play a role in the further discourse. An example is, for instance, shifting from $\frac{1}{2}$, considered as a single mathematical object to $\frac{1}{2}$ as the representative of an equivalence class in order to anticipate an operation te be performed with $\frac{1}{2}$.

The other types of shifting will be illustrated frequently.

I return to the course of fractions.

What happened with the design in this respect?

What happens is that attention is shifted from the table (A) to the kitchen

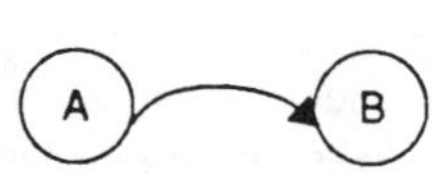

Fig. 7. Parallel shifting (ps).

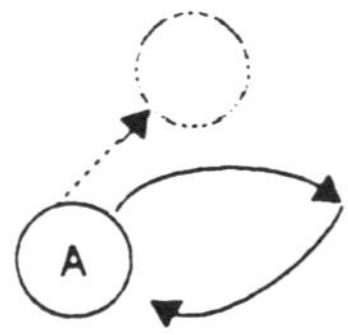

Fig. 8. Reflexive shifting (rxs).

(B), so that through the aspect of serving the division at the table (A) can be illustrated even better. By also redirecting the emphasis from the bars, pizzas, etc. to the dividers themselves, with in mind the control of the situation, – 'reciprocal shifting' occurs once again. Moreover the designer is faced with the opportunity to interweave fractions with ratios, the more so because it is evident – in anticipation of fair sharing – to already compare such divide situations (according to their ratios).

The shift in emphasis reveals a junction of the interweaving of learning strands, namely of fractions with ratios. Because subjects of learning material are at issue here which each demand their own learning courses, I will call this external interweaving. This shift also makes it clear that as a designer, one can in such a case choose the direction of mathematizing, namely the track of ratios, of fractions or of both.

In this case the junction of interweaving can be reinforced considerably by incorporating the comparison of situations, because one then also anticipates on the determining of the difference of fractions with different denominators. (Streefland, 1991a).

For example, $\frac{\text{(3b)}}{4}$ can – in first instance – be compared with $\frac{\text{(4b)}}{5}$ without already knowing the exact portions. And that is precisely what ratios are for (Freudenthal, 1983, Ch. 6). Comparisons like this can bring about a considerable 'shift' in the reasoning of the pupils. What it comes down to is $\frac{\text{(3b)}}{4}$ (A). To be able to draw conclusions about it, $\frac{\text{(4b)}}{5}$ (B) is placed along side. This provokes the transition from A to B to (also) be able to conclude something about A, reciprocal shifting therefore. But shifting of the attention in the situations themselves to their complementary situations $\frac{\text{(1b)}}{4}$ and $\frac{\text{(1b)}}{5}$, in order to consider the comparison of the original situations, is also incorporated, hence directed shifting. Depending on whether one carries on with these newly determined situations or goes back to the original situation there is evidence of both forms of shifting. In any case there is evidence of a learning activity that is rich in possibilities of changing perspective.

I feel that thus I have been able to illustrate that at issue is a complex junction of learning strands, rich in possibilities to proceed in different directions. Developmental research, followed by comparison of possible alternative prototypes of learning courses can render a conclusion about one or more desired directions. In traditional didactics this junction went unnoticed entirely, because the approach to the area took place on a too formal level. Better even to say that the junctions are not made until the ends of learning strands. I will return to this matter.

One and other makes it clear that

– catching the exact degree of concreteness

is a most important point of attention for a designer. Or in other words, to reveal the junctions for interweaving of learning strands with different possibilities for mathematical progression one takes an unbiased look at a context. And one observes the degree of mathematizing, if there is (already) evidence of the same. As the age of the pupil for whom a subject is intended becomes higher, one will have to take this into account to a greater degree. Also then will a qualitative and estimated approach retain didactical value. One will however sooner be able to place the pupils before more advanced levels of mathematizing. Their arsenal of mathematical tools has gradually expanded as the learning process progresses.

Notation, Schemes and Models

The disclosed possibilities for external interweaving of learning strands can provide a fruitful spin off as far as the teaching learning process is concerned. The model situation of the seating arrangement, materialized in symbol and scheme, becomes a context or situation model.

Moreover this provides the starting point of a process of progressing schematisation, steered by this situation model. There is also a close tie with other activities in the course, for instance with dividing.

The looking back from a (constituent) portion to situations of origin is a genuine change in perspective. The information of the division situation have been exchanged against the outcome thereof, resulting in an idea for the design of a good (re)construction problem:

$\frac{1}{4}$b ⟶

1b	2b	3b	
4	8	12	

etc.

The importance of this looking back to the sources of origin of the describers of part-whole relations is two-fold. Not only is the contextual meaning of such relations reconfirmed. At the same time one anticipates on the mathematisation of equivalence, i.e. on the formalising of this phenomenon. Hereby the contents is as it were, taken along for the moment, the exchange of content for form is postponed for a time. However, the ratio table stems automatically, in some form or other, from this activity. This model scheme *of* the organisation of division situations of equal portions can in future be turned into a model scheme *for* ratio in general. Or, in other words, this scheme as post-image of mathematizing of division situations from now on becomes a pre-image for the mathematizing of ratio in general. By now it will be clear that this will (have to) go hand in hand with a marked change in standpoint, i.e. directed shifting.

A notable side effect is the possibility for progressing schematisation which the ratio table also has and to which I have referred several times in a different context (see for example Streefland, 1982 and 1991a).

As far as the mini-monographs are concerned, one sees the same – on a broader scale – in first instance as in looking back from portion to situation. But there is more going on! Suppose this concerns telling a 'story' about $\frac{5}{6}$ (bar). Again there is evidence of change in perspective. The consequence is a (re)construction problem. Making the connection between $\frac{5}{6}b$ with $\frac{\bigcirc}{6}$ can evoke ideas about per item and otherwise dividing. $\frac{5}{6}b = \frac{1}{6}b + \frac{1}{6}b + \frac{1}{6}b + \frac{1}{6}b + \frac{1}{6}b$ and $\frac{5}{6}b = \frac{1}{2}b + \frac{1}{3}b$ (first dividing 3 bars and then 2). The production of relationships is in first instance based on the different manners in which the division occurs. While doing so, a change can however take place in the pupils, namely from the composing of relations on the basis of concrete division to the organisation of the (fraction) material in the relationships themselves.

Example:

$$
\begin{array}{ll}
\frac{5}{6}b = \frac{1}{6}b + \frac{1}{6}b + \frac{1}{6}b + \frac{1}{6}b + \frac{1}{6}b \rightarrow & = \ 5 \times \frac{1}{6}b \\
\quad = \frac{3}{6}b + \frac{2}{6}b & = \ 2\frac{1}{2}b \times \frac{1}{3}b \ (^{*}) \\
\quad = \frac{2}{6}b + \frac{3}{6}b \text{ (comm).} & = 10 \times \frac{1}{12}b \ (^{*}) \\
\quad = \frac{1}{3}b + \frac{1}{2}b \text{ (equiv.)} & \cdots\cdots\cdots\cdots \\
\quad \cdots\cdots\cdots\cdots\cdots\cdots & \text{etc.}
\end{array}
$$

(*) (doubling/halving of factors)

The 'shifting' process is characterised by 'direct' and 'reciprocal' components. Instead of descriptive material of division and the portions, the real fractions are gradually extended to an algebraic system with its properties, rules and laws. In essence it comes down to a complex form of 'directed shifting'. By the change in perspective which the designer applies, he elicits reflection by the pupils. Their attention displaces itself from the concrete division situations (A) to the fraction material itself (B) with in mind the production of a mini-monograph (C). What in first instance is a (re)construction problem matures into a problem of (free) production. This is revealed by the change of perspective of the designer. In developmental research 'directed shifting' was detected in the learning process of all participating pupils (Streefland, 1988; 1991a).

Again it is a shift in the perspective of the designer which brings to light a junction of interweaving of learning strands, but of a different nature than in the earlier mentioned case. By construing mini-monographs the pupils get to practice all kinds of skills. Not only as such, but serving the higher objective of vertical mathematizing, the algebrising of the system of fractions. This occurs in close connection with the concrete sources of insight. The question which the designer is faced with is how to interweave this practising with the learning strand of the forming of concept itself. (Hence that I will call this a junction of *internal* interweaving). And the answer has actually already been given. The outlined (re)construction problem comes to light every time the designer applies the mentioned shift in perspective. This problem has a place in the line of the development of concept as well as in the line of vertical mathematizing.

Again there is evidence of looking back with a view to anticipate. Important activities in this connection are learning to relate given mini- monographs with concrete (division) situations and the explanation of the origin and relationship within and between mini-monographs (Which production rules are applied?), a good example for interactivity in the classroom after the own production of the pupils (Streefland, 1988; 1991a).

Notation on the Long Term

In this criticism of mathematics as propagated by 'new-Math' Freudenthal (1973) employed the concept of antididactical inversion. By this he meant that the 'smoothed over' end product of the historical learning process was the point of departure in education. Beginning with definitions, axioms and other formal finesses, in short the genesis in reverse as a didactical principle.

The movement of realistic mathematics instruction can be considered as the reaction to this practice. It is an example of directed shifting on a very global level. For the designer of realistic mathematics instruction this has the consequence however that the interweaving of learning strands, e.g. for ratio and fractions, must already be put into effect at an early stage of the teaching process, as I have shown (also see Streefland, 1984b). Something like this can most definitely not be postponed until the particular subjects (for external interweaving) or else parts of the course (for internal interweaving) have already been abstracted to the stage that only the formal connections based on algorithms, laws and rules can be exacted from the material. Examples are the mathematical identification of the equivalence of ratios and fractions and their conversion into ones with the same 'denominator'. An approach such as this proved unsuitable in traditional education as well.

With in mind the maintaining of content while postponing the more formal, there has recently been argument in favour of preserving the origin of the fraction in terms of natural magnitudes, in the notation (Treffers, 1991). This means that there was a search for strong 'models of...' which would not lose in power as 'models for...'. These models are then coupled to a notation in which the natural quantities which give the divided and measured objects their value (length, price, duration,...) are still present: $\frac{1}{2}$ meter, $\frac{1}{4}$ guilder, $\frac{1}{3}$ hour, etc. At issue again is 'directed shifting', both at the level of the lesson as well as the learning course, but now in opposite direction as in which Freudenthal's antididactical inversion worked.

The shifting of attention to the visualisation and the magnitudes, present in the objects, does not only set model forming into motion, time and again, but also gives to the fraction as describer of relations its new status of operator, as I have shown. By retaining the natural quantities in the notation the content of the fractions are taken along on the path to formalisation. The antididactical inversion is completely cancelled out in this way. And what is more, abstraction (and generalisation) is placed in a new light (also see Van den Brink, 1989).

Suppose that the children have learned what $\frac{1}{2}b$, $\frac{1}{4}b$, $\frac{1}{2}b + \frac{1}{4}b$ etc. is, with $b =$bar. Now 100 grammes, 200 cents, etc. is substituted respectively for b. In the terms of Van den Brink (1989) the pupils are given the opportunity to 'decorate'

$\frac{1}{2}b + \frac{1}{2}b$ etc. with new, context restricted meanings. One sees that $\frac{1}{2}b + \frac{1}{4}b$ etc. will not start behaving differently with these new 'decorations'. They stay themselves permanently and thus in time become what they are to the pupils, i.e., formal fractions, mathematical objects, mental objects or noumena, gradually freed from their context-restrictedness, and operational.

Thus the process of abstraction not by taking away from, but rather by adding to, is a significant example of 'parallel shifting'; the shifting of (context)environment to environment.

Each time other context environments are assigned to the same models, notations and operations, so that ultimately a formal fraction construction results, also in the mind of the learning individual. They have then come to (formal) understanding and have left the stage of the mental objects behind them entirely. This is in part due to operating on the bypaths of the entire quantity values, a form of 'reciprocal shifting'. Gradually the uncommon denominators of the fractions will be a focal point in the activities of the pupils, at the expense of the values which they have in the natural quantities. With growing insight these will be attuned accurately and easily according to the lcm-principle.

Example:

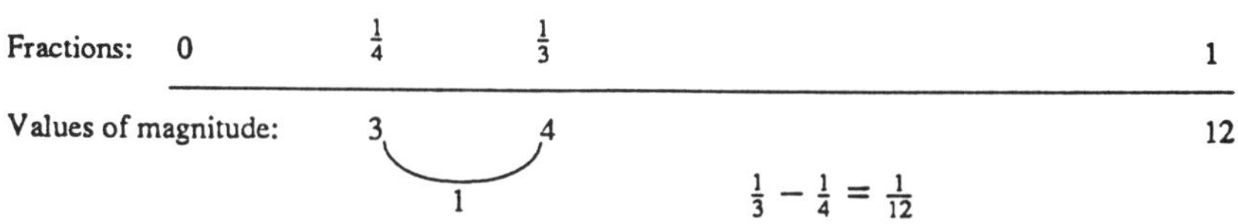

The pupils will have to shift, whether they wish to or not, namely a complex compilation of the three already distinguished forms of shifting, in consequence of the strategy applied by the designer. This reflects the complicated dynamics of the learning activities that are elicited by such a thought process.

PRELIMINARY BALANCE

A designer has at his disposal a repertory of great strategies. The application thereof largely determines the architecture of a learning course under development. They come down to:

– catching the right degree of precision, as approach strategy

In this connection a 'qualitative approach' and 'estimation' were mentioned. More in particular this concerned a balancing of activities thus that the pursued insight cannot be obstructed by premature rules (soles of the giant).

Remark: Naturally this is not the only approach strategy possible. One can, for example, also choose to bring to light the informal knowledge which pupils have about a certain subject, for example by confronting them with relevant (conflict) situations. Strategies related to the foregoing are:

– catching the right degree of concreteness, or
– catching the right stage in the interweaving of learning strands, or

– catching the right level of mathematizing.

Proven to bear fruit is:

– the application of different variations of shifting thus to exploit a context to the full. It served to trace a junction of external interweaving of learning strands (ratios and fractions). This was substantially reinforced by also bringing in comparison of (division) situations to the junction.

Of inestimable value – together with the distinguished forms of shifting – is the strategy of change in perspective or standpoint. It is – perhaps- the most productive and influential of all. Change of perspective can be distinguished from the mentioned forms of shifting by exchanging a part of the information in the problem situation for that which is being asked. On the basis of the extent of the exchange we can speak of *partial* or *complete* exchange of perspective (Streefland, 1992). Possibilities for (re)construction and production problems become evident in consequence. Moreover, as a result of the shift which occurs in the mental activities of the pupils, (re)construction problems prove to turn into true production problems (mini-monographs). True free production problems (e.g. making a test on one's own by looking back on a topic or course) rest on a complete change in perspective. All of the information is left aside. Only that which is asked (the test about the given subject) remains.

This reveals yet another distinction between change in perspective and shifting. Change in perspective is employed as a strategy mainly there where some degree of mathematizing is already taking place. There is information (the fruits of mathematizing) which can be exchanged for a question (towards further mathematizing). The shifting is well suited for explorative operations in the context. However, application of the shifting on material in which vertical mathematizing is already in progress, is not of course precluded, as I have demonstrated in connection with abstraction and will yet show in another connection. In these cases structuring a course globally is involved.

Having come this far, the very important first strategy, which I have saved for this moment, can be mentioned, namely:

– *catching the right (mathematical– didactical) context.*

Now that there are strategies by which to test a context on its promise, it can be examined whether as a source it is meaningful and rich enough in terms of mathematizing, not only in the manner of approach, but also for the longer term of the pursued learning process. The context of the restaurant with its possibilities for serving, seating arrangements and division, proves to meet these criteria, as does the item about the busses, which will be shown later.

In connection with shifting and changing perspective there is one more important conclusion for design.

The situating of (re)construction and (free)production problems in the alignment of a learning course is no longer a question of the personal taste of the designer. Their place is – as we have just seen – much more determined by the contexts and activities which the designer employs. He detects them by applying the strategy of changing perspective. They play their role in the various stages of the mathematizing process. And because we are talking about a shift in a certain

Fig. 9. The bus arrives at the bus stop with 3 passengers on board etc.

problem one will in general introduce such a construction or production problem with sufficient direct questioning.

These (re)construction and production problems in turn are junction of internal interweaving for the learning strands of practising skills (on the road to formalising) and the forming of conception. The pursued raising of levels is incorporated herein, both because of the principle of looking back in order to anticipate as well as because of the changes in standpoint which can be pointed out in the learning course and are demonstrably elicited from the pupils (Streefland, 1991a).

In conclusion, the application of changing the perspective brings with it some important tools for mathematizing, such as the forming of notation, schemes and (situation) models, which are essential for the further design of a learning course.

EXCHANGE OF PERSPECTIVE WIDELY ACTIVE

Examples from learning courses for addition and substraction, the counting of large quantities, percentages and sine functions illustrate the great strength of change in perspective as a strategy for the revelation of junctions of interweaving learning strands.

First Example: The Bus

The getting on and off of passengers at a bus stop has proven to be a suitable context for addition and subtraction. In the latest generation of textbooks in The Netherlands this context is used for that purpose (Figure 9). Attention is directed at what goes on at the bus stop (Van den Brink, 1989).

Remark: With in mind the application of shifting on this context, attention could also be directed at that which takes place inside the bus: standing or sitting, sitting in the back or up front, or – in the case of a double- decker – sitting upstairs or downstairs. A number of until now partially ignored possibilities for the development of the budding concept of numbers presents itself.

Back to the bus stop. The exchange of bus stop-given and occupation of the bus after it leaves the bus stop renders an interesting variant (Figure 10). The events which take place at the bus stop can be described with number pairs, (0 get on, 3 get off), (1, 4), 2,5), (3,6),...

All these pairs describe '3 less' passengers after the bus stop. A junction of external interweaving for the initial addition and subtraction with negative numbers in pairs has emerged from the applied partial change in perspective.

Fig. 10. What happened at the bus stop?

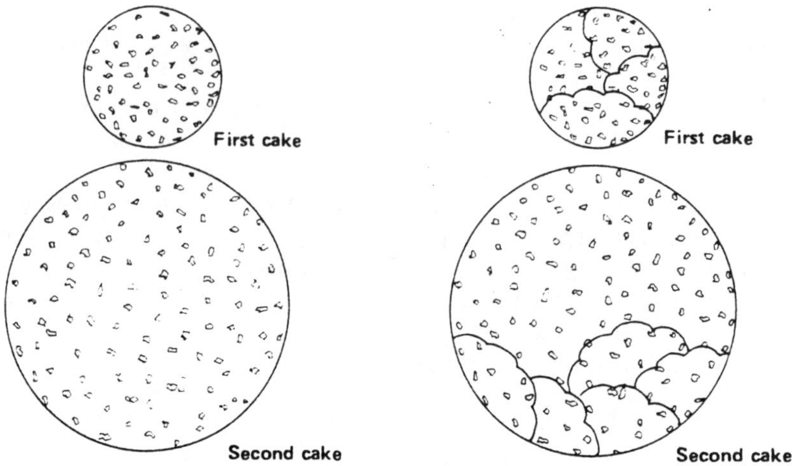

Fig. 11. Ginger cakes.

If the change in perspective is made more complete by also leaving the occupation of the bus before the bus stop unmentioned, then a production problem with the corresponding possibilities is the result.

Second Example: Taste of Ginger

Giant Egbert from the earlier mentioned topic was given a ginger cake. The first question (Figure 11) deals with the recipes for two ginger cakes in which one is four times as large as the other.

a. Which cake contains the most chunks of ginger? (e.g. 60 and 120 bits respectively). The results of counting are next related to the taste of the cakes.

b. Which cake has the most bits of ginger per bite? Here the change in perspective – reciprocal shifting in this case – does not only reveal a junction of external interweaving (large numbers and density). The children must also settle a conflict, namely that absolute and relative consideration of the counted amounts (seems to) conflict.

Third Example: Copying Machine

In a learning course on percentages there is the exploration of a copying machine (Streefland, *et al.*, 1991). An original is placed on the copier. The image appears

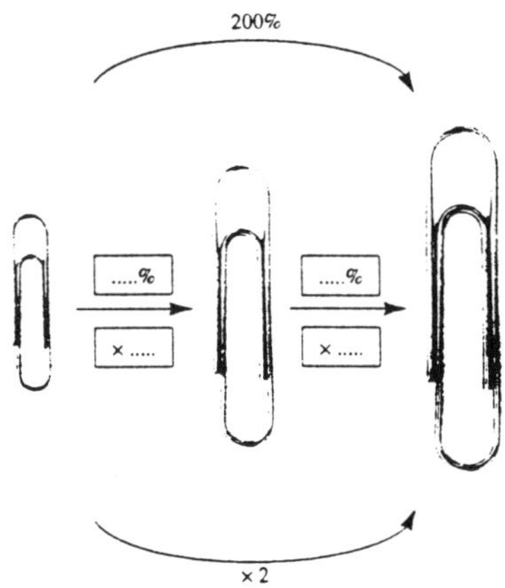

Fig. 12. Copying machine.

to be twice as large (linearly) (Figure 12). The copier was set at 200%. If the set percentage is not given and it is said that the enlargement with a factor two was achieved in two times, which means that one has asked for a copier setting of 141%, then one interweaves percentages with (exponential) growth, and therefore with taking square roots and raising to a higher power. $1.41 \times 1.41 = 2$ (a good example for exploration with the aid of an electronic calculator). Again a change in perspective proves to reveal a junction of external interweaving of learning strands.

Fourth Example: Sine Functions

In a learning strand for sine functions it is not uncommon to devote attention to the attainment of skills such as:

- the drawing of a sinusoid for a given formula, and to
- finding a formula for a given graph (Figure 13) (De Lange and Kindt, 1984; Kindt, 1989).

These are results of changes in perspective comparable to open number problems from traditional arithmetic. There is however a much more productive alternative.

Suppose, a graph is given, referring to the sources of origin of sine functions (waterwheel with lighted point, Ferris wheel at the fair, lighted point on turning propeller of a flying aircraft, harmonic oscillation, Figure 14).

The amplitude and period can be according to choice under condition that justice is done to their relationship, i.e. under this restrictive conditions the coordinate system is optional. E.g. Determining the amplitude at 1 and therefore the period at π (because of their ratio) and the basic level at 1, loads to Figure 15. If an amplitude of $\frac{1}{2}$ had been chosen however, with $\frac{1}{2}\pi$ for the period and the same basic level, we would end up with Figure 16. In this manner 'families' of sine functions can be generated by looking back at the sources of insight

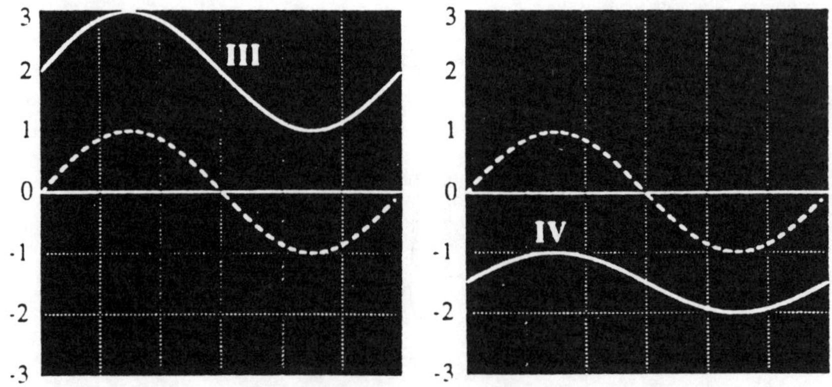

Fig. 13. Finding a formula.

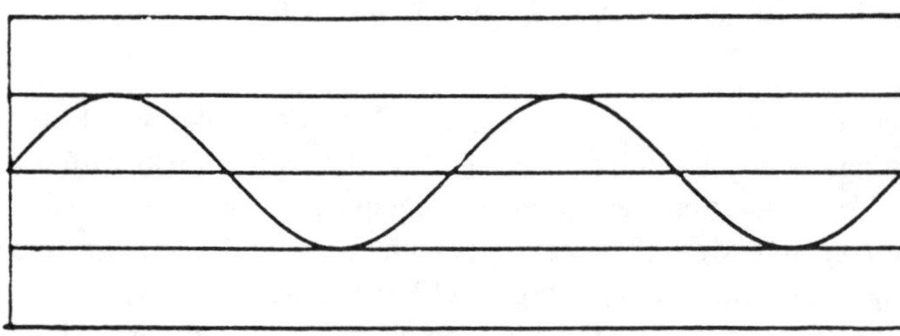

Fig. 14. Which is sinusoid?

while changing around the parameters. The skill of drawing up formulas will be developed as a result. The practice does not only have value in itself but also serves a furtherreaching objective.

The produced material can be organised at even a higher level of abstraction and generalisation, expressed in formulas of the type $y = a \sin bx + c$. The applied, complete change in perspective does not only lead to an assignment of free production. At the same time the junction of internal interweaving appears, namely for the practice (serving progressive mathematizing) with the acquisition of comprehension. Important operations such as generalising by unifying become manifest.

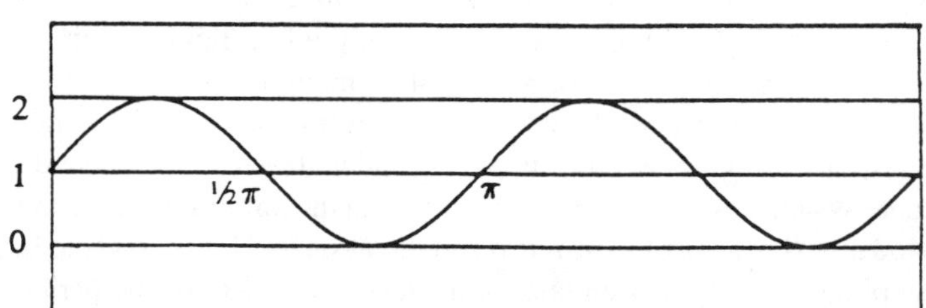

Fig. 15. $y = \sin 2x + 1$.

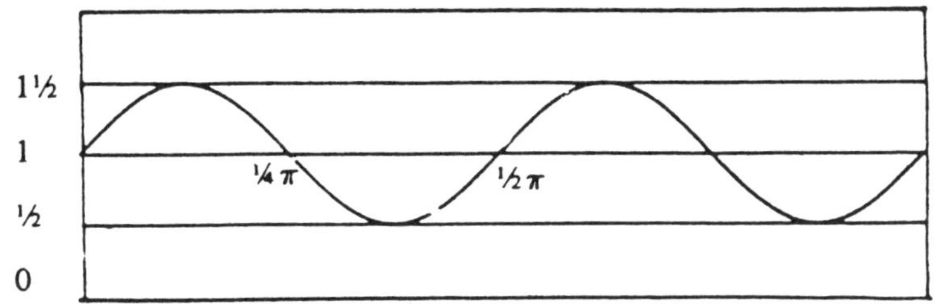

Fig. 16. $y = \frac{1}{2}\sin 4x + 1$.

Side step

While we are on the subject, this is a good opportunity to touch on yet another strategy for the designer. Let me return to Figure 13.

The basic model of the sine function is found dotted in and the graphs of two other sine functions are indicated. Let us suppose that the question reads: How can graph IV be derived from the basic model? (What has been changed in the dotted graph to get graph IV? What has happened?) The pupils can be allowed to respond in terms such as: The tops of the waves are lower and the dips more shallow and the water level has dropped (amplitude and basic level). Or: The graph has been pressed down somewhat and flattened out. Instead of such relative, informal language one can – in first instance – even accept pure pointing out: the graph has been shifted from here to there and...

There are great advantages in allowing pupils to describe the changes in a qualitative sense first. The attention is fixed on – in this case – amplitude and basic level, but in consequence also on that which remains unchanged (the distance between the points where the wave tops turn into wave dips and their place, period, frequency).

This grammatical contribution by the pupils can now be developed to good use. The posing of corresponding questions can be left to the pupils themselves. By how much do the wave tops become lower and the dips more shallow? How much did the basic level drop? What remained unchanged? In a most self evident manner the answers lead to $y = \frac{1}{2}\sin x - 1\frac{1}{2}$. That concludes the intended strategy for the present. It comes down to:

– the development of math language at a level that surpasses the levels of the demonstrative and linguistic relative, in particular at the level of conventional variables and functional description. It does not need saying that in an activity of this kind change in perspective in the form of shifting plays a role.

Closely related to this strategy is also the following:

– development of representations, forming of notation and symbol language in such a manner that this gradually leads down the path from content to form, preventing preliminary violation of either.

For example: Representing the result of a division respectively by: 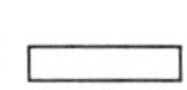 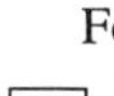, a whole bar and another piece, a whole bar and another half, one and a half bar, 1 and $\frac{1}{2}$ bar, $1b + \frac{1}{2}b$, $1\frac{1}{2}b$, $1\frac{1}{2}$.

THEORETICAL DESIGN OF REALISTIC MATH EDUCATION

Of the major strategies that a designer has at his disposal the most important one has thus far not been mentioned, namely reflection on own action and reasoning and – of course – on that of others. The foregoing paragraphs did however contain the fruits of such reflection. But the end is not yet in sight. The question that now becomes important concerns the consequences of the foregoing for the status of the theoretical framework for realistic math education. This framework as a descriptive framework has clearly become dated.

Earlier on I mentioned the level thereof concretely in a metatheoretical sense. The possibilities have now been indicated to rise beyond this. We have seen that reality as a source and application area forms the basic level par excellence for realistic mathematics instruction. The revealed strategies can be applied as *operations* on a selected reality with its phenomena. From this comes notation forming, schemes and models, problems of (re)construction and (free) production and various kinds of junctions of interweaving of learning strands.

(It is unavoidable however that still the designer is urged to display his experience, intuition and creativity; see in this connection Gravemeijer, 1987). Whatever the case, possible versions for the architecture of a learning course are not only made considerably clearer in this manner, but their global structure is revealed in full. Nevertheless it will continue to be necessary to make choices, more frequently even as the didactical fine tuning of the design increases and the activities are ever more directed towards the local content of separate activities or lessons. Hence that developmental research will always be necessary in regard to variations of prototypes for a learning course, possibly concluded with comparative research on the effects thereof (see Freudenthal, 1991, 3.2.3. p. 156).

The yield of the 'operations' which are applied to realistic contexts illustrates how levels are and can be raised. And the designer anticipates in this manner on the progression of the interaction in the instruction that is to be realised. His strategies can also be used there. Giving shape to interactive teaching on the grounds of that which the pupils construct, produce, etc. is also a matter of designing. And the (preliminary) design of the designer anticipates on this contribution.

The thought experiment of the designer becomes reality in the interaction. And that which is unintentional and does not work is also revealed. And also who in fact produces the mathematics, i.e. whether the teacher succeeds in giving his pupils the necessary (construction) space and knows how to deal with it (Streefland and Te Woerd, 1991). If the teacher succeeds in catching the right degree of concreteness, the right degree of precision, the level of mathematizing etc of the material contributed by the pupils, then the process of interaction can be built up on the foundation of level theoretical considerations, by placing the emphasis in discussions of and negotiations about material on the transition of levels which are inherent to that material (Streefland, 1991a, Chapters 6 and 8).

The theory available thus far therefore proves to be twice 'nested'. Reality as a source and application area encompasses all other basic characteristics of learning material nature, at least has that potential. In turn this is dissolved wholly in the interaction. These conclusions originate through global application of directed shifting. The revealed strategies act as the operational relationships between the various building blocks of realistic theory as it served as description framework thus far (also see Streefland, 1991b). The theory is thereby lifted to the two next higher levels above the concrete. For this second and notably third level consider for example the connection between 'context' and 'shifting' and between 'shifting' and 'change in perspective' which was discussed.

Can realistic theory now adopt this as genuine design theory? There is no doubt that a step has been made in the right direction, but one could retort that the required mathematical-didactical attitude, characterised by the unfolded strategies, and in which the similarity with a mathematical attitude becomes manifest (Freudenthal, 1991) need not be part of the (conscious) repertory of a designer. For this the certainty is lacking that the chosen contexts are the right ones and especially, that they will be processed properly, while – it has been shown – only the proper kind of processing can reveal the mathematical-didactical richness of a context. As such this is correct. I have yet to make the final step. This step opens up the path to theoretical design to its very end. The strategies or processes to be made for the designing were made deliberately. Hence they can become a tactic, i.e. one of the rules of design. To illustrate, a few more small examples of context analyses, aimed at *making an inventory* of standpoints.

More Busses

The activities at the bus stop have not been utilised to the full. Besides attention for getting on and off, one can also focus on transferring and the changes that means for the number of passengers at the bus stop itself, in the event there are several bus lines. This again opens new informal approaches to addition and subtraction.

Banking

From the viewpoint of the account holder one frequently introduces negative figures, balances and deficits. If the banker's point of view is also taken into account, one will soon discover that much richter material for negative numbers can be elicited from this context. Then the 'bookkeeping' becomes two-way, giving cause for much reflection by the pupils.

Sine Functions

Let us take one more look at the examples of the turning waterwheel with a lighted point (height of the point in relation to the water level) and of the aircraft with its turning propeller (height of the point in relation to the shaft). In both cases the shifting of attention, namely from (horizontal) stream to (vertical) waterfall, from side view to view from above, appear to reveal the internal interweaving of the sine with the cosine and moreover that the last $\pi/2$ anticipates on the first, whereby a formula like $\cos t = \sin(t + \pi/2)$ suddenly becomes very meaningful.

The results of a suchlike analysis can thereafter be used in design. The finding of these standpoints no longer depends on the fortunate bright idea of the designer through change in perspective or shifting. It is the deliberate preliminary analysis which divulges all the possible standpoints in a context for a design. Something similar also applies for the other strategies that were mentioned. Once conscious of them they can be elevated to a method of operation.

From all of this, realistic theory has attained a higher theoretical status and as design theory has gone from post-image to pre-image. For those that pursue the design of realistic mathematics instruction this theory can now finally be set as a standard, because the question of the 'how' of its application has been answered to the fullest degree.

CONCLUSION

Realistic theory has not yet crystallised out by any means. There has been considerable progress. The meaning and extent of this progress will become visible on the long term. Many new experiences with design and developmental research will be necessary first. But at any rate a change has been brought about in how one looks at (the design of) realistic mathematics instruction. A change in perspective is essential therein, as is the interweaving of learning strands.

The frequent application of forms of notation, schemes and models, the making of own constructions and productions together with practice, elevate themselves in terms of theoretical status above reality as a source and application area, as I have shown. This is not only because the characteristics as mentioned can be derived from the realistic context through the unfolded operations, but also in view of their function in progressing mathematisation. What really matters in designing a course is anticipating to the intended learning process by interweaving the mentioned topics, either with the realistic source (consider in this connection the new light into which abstraction was placed), or with the realistic application area. For each next step in the design the outlined operations become available once more, so that having arrived there, the junctions of interweaving of learning strands can again be revealed, or at least relinquish the (re)construction and production moments.

About change of perspective as a strategy for the solving of problems Freudenthal (1991, p. 40) wrote:

> I didn't pay enough attention to change of perspective. ..... The subject deserves a more systematic treatment, which I do not dare undertake. (Freudenthal, 1991, p. 41)

Does that mean that this writer could be that presumptuous, or even suggest that he did dare to undertake such systematic treatment? Or even to have achieved that? Nothing is further from the truth. But... and in this connection I quote another statement from Freudenthal that I have quoted before (Streefland, 1991b, p. 11):

> Learning processes are marked by a succession of changes of perspective which should be provoked and reinforced by those who are expected to guide them. (Freudenthal, 1991, p. 94)

The analysis of a learning course for fractions illustrated how the manner of design can provoke and reinforce such a change in perspectives in the pupil. Namely by employing this same method in the didactical activity of designing. With the major strategies of didactising we afford ourselves access to the art of designing.

In the developmental research that I have conducted (Streefland, 1991a) the didactical reality of an instruction experiment served to illustrate the role of change of perspective and other strategies in the spirit of Freudenthal's last quote. It appeared – and analysis of the learning course bears this out – that reflection can be active much more widely than suggested by theory thus far. As appears this metacognitive activity not only occurs when executing (free) production assignments or the settling of (cognitive) conflicts.

One can call the entire process of developmental research horizontal didactising (Freudenthal, 1991, p. 45). In the design of a long learning strand for sine functions the paradigmatising or vertical didactising took place (Streefland, 1992) as did the adjustment of the learning course design for fractions on the basis of suggestions from Treffers (Streefland, 1991a). Becoming aware of the theoretical implications form the (preliminary) crown of these endeavours.

The outlined theoretical breakthrough, or rather discovery, can be compared to the discovery of the level theory by the Van Hiele's (see Freudenthal, 1991, p. 102). The material on which the Van Hiele's founded and applied their theory was considerably more surveyable and modest in size. The conditions on the grounds of available preliminary work were however much less in their favour.

It is of course not the case that the ideas which are unfolded here deserve the qualification of 'new' in the sense that they are indeed brand new, as granted by deus ex machina. I would by do means make such a claim.

In the course of the years my work has confirmed especially many of the ideas of Freudenthal. On the other hand this is 'new' in the sense of a 'new outlook' because here a number of theoretical connections have been made which have brought the organisation of the available knowledge to a higher level.

To employ a physical metaphor, which I would borrow from Van Est (1991) for this occasion, in Freudenthal's work and that of others and of my own the unfolded ideas were already present in a saturated solution. What I have done is crystallize them out. Yet it was a long time coming, because notably the issue of the interweaving of learning strands was operational theoretically speaking in an early stage (Streefland, 1984a, 1984a). Perhaps too early on. At any rate, the operational possibilities remained latent for a long time. Perhaps the time was just not ripe. Thanks to Freudenthal's posthumous 'Revisiting Mathematics Education' I picked up the thread once more. He was the one who in the last year of his life, not only through this book but also on a number of other occasions, returned me to the track of earlier ideas. It is for more than this that I am grateful to him.

REFERENCES

Brink, F.J. van den: 1989, *Realistisch rekenonderwijs aan jonge kinderen*, RUU, OW & OC, Utrecht.

Casimir, H.B.G.: 1987, *Waarheid en Visie Over Wetenschap en Maatschappij*, Meulenhoff, Amsterdam.

Est, W.T. van: 1991, 'Hans Freudenthal', *Jaarboek 1991 van de Koninklijke Academie van Wetenschappen*, 139–142.

Freudenthal, H.: 1968, 'Why to teach mathematics so as to be useful', *Educational Studies in Mathematics* **1**, 3–8.

Freudenthal, H.: 1973, *Mathematics as an Educational Task*, D. Reidel Publ. Co., Dordrecht.

Freudenthal, H.: 1979, 'How does reflective thinking develop?', *Proceedings of the third International Conference for the Psychology of Mathematics Education*, Warwick.

Freudenthal, H.: 1981, 'Major Problems of Mathematics Education', *Educational Studies in Mathematics* **12**, 133–150.

Freudenthal, H.: 1983, *Didactical Phenomenology of Mathematical Structures*, D. Reidel Publ. Co., Dordrecht.

Freudenthal, H.: 1987, 'Theorievorming bij het wiskundeonderwijs. Geraamte en gereedschap', *Tijdschrift voor nascholing en onderzoek van het rekenwiskundeonderwijs* **5**,(3), 4–15.

Freudenthal, H., 1991, *Revisiting Mathematics Education: China Lectures*, Kluwer, Acad. Publ., Dordrecht.

Goffree, F. and Stroomberg, H.: 1991, *De educatieve ontwerper – productief, reflectief, studieus*, SLO, UvA, Enschede, Amsterdam, (draft version).

Gravemeijer, K.: 1987, 'Three Dimensions – een model voor doel- en theoriebeschrijving', *Tijdschrift voor nascholing en onderzoek van het rekenwiskundeonderwijs* **5**,(3), 46–55.

Jong, R.A. de, (ed.): 1978, *Oppervlakte (2)*, Leerplanpublikatie 9, IOWO, Utrecht.

Kindt, M.: 1989, *Sinus en Co*, OW & OC, Utrecht.

Kühnel, J.: 1966, *Neubau des Rechenunterrichts*, Düsseldorf, (first edition, 1916).

Lange, J. de and Kindt, M.: 1984, *Sinus*, Educaboek, Culemborg.

Streefland, L.: 1982, 'Subtracting Fractions with Different Denominators', *Educational Studies in Mathematics* **13**, 233–255.

Streefland, L.: 1984a, 'Search for the roots of ratio: Some thoughts on the long term learning process (Towards ... a theory). Part I: Reflections on a teaching experiment', *Educational Studies in Mathematics* **15**, 327–348.

Streefland, L.: 1984b, 'Dichtheden in leerprocessen', *Tijdschrift voor Didactiek der Natuurwetenschappen* **2**(2), 72–87.

Streefland, L.: 1985, 'Wiskunde als activiteit en de realiteit als bron', *Nieuwe Wiskrant* **5**(1), 60–68.

Streefland, L.: 1988, 'Reconstructive Learning', *Proceedings of the 12th International Conference Psychology of Mathematics Education*, Veszprém, Hungary, Vol. I, 75–92.

Streefland, L.: 1991a, *Fractions in Realistic Mathematics Education. A Paradigm of Developmental Research*, Kluwer Acad Publ., Dordrecht.

Streefland, L.: 1991b, 'Het toetsen van denkstrategieën in heuristisch wiskundeonderwijs. Kan dat? Hoe? *Tijdschrift voor nascholing en onderzoek van het rekenwiskundeonderwijs* **10**(1), 3–21.

Streefland, L.: 1992, 'Sinusfuncties in ontwikkelingsonderzoek. Een gedachtenexperiment', *Tijdschrift voor Didactiek der Natuurwetenschappen* **10**(1), in press.

Streefland, L., Moor, E. de, Treffers, A.: 1991, 'Proeve van een nationaal programma voor het rekenwiskundeonderwijs op de basisschool (14); procenten – didactische notities, *Tijdschrift voor nascholing en onderzoek van het rekenwiskundeonderwijs* **10**(1), 29–39.

Streefland, L. and Woerd, E. te: 1991, *Protocolanalyse kwalitatief.* Een in het kader van het More-onderzoek uitgevoerde kwalitatieve analyse van lesprotocollen, Utrecht, Freudenthal Institute, (in press).

Treffers, A.: 1987, *Three Dimensions. A Model of Goal and Theory Description in Mathematics Instruction – The Wiskobas Project*, D. Reidel Publ. Co. Dordrecht.
Treffers, A.: 1991a, 'Didactical background of a mathematics program for primary education', in Streefland, L. (ed.), *Realistic Mathematics Education in Primary School: On the Occasion of the opening of the Freudenthal Institute*, Freudenthal Institute/CD-β, Utrecht, pp. 21–57.
Treffers, A.: 1991b, 'Breuken op de dubbele getallenlijn', *Tijdschrift voor nascholing en onderzoek van het rekenwiskundeonderwijs* **9**(4), 2–14.

Freudenthal Institute,
Utrecht University
Utrecht,
The Netherlands

(H.F. in: Editorial letter 12 august 1978):

For me this is the essential point. This is not at all the way in which to understand problem solving, heuristics, nor for that matter, teaching. It is also not justified to characterize it as a science as compared to art. If it is a science, it is an artless one.

My remarks on the subject of teaching integrals, to which you reacted, are secondary. This does not mean they are unimportant. Your comparison with cancer research in rats and human beings is mistaken. If you do research on heuristics and problem solving, you may indeed choose any subject area where such activities are undertaken. The intrinsic value of a subject can be an indication of such possibilities. Integration, as it was understood fifty years ago almost everywhere and still is at some institutions of teaching today is a subject matter that derives its importance from being a source for test problems, solved by means of a large number of clues to be recalled and applied if necessary. It is exactly the opposite with heuristics and problem solving. This is how the value of subject matter appears to me.

Finally, what matters is not the writing of a paper that is 'more consistent with my taste'. Many papers were published in ESM that were not my taste at all. I allowed many papers to be published that I personally did not like. There are tastes that I do not share and I try to respect them. It is the same as in mathematics itself. There are subjects and methods that I have never liked but I would never dare say that they are bad mathematics. Education is different. There is certainly no unanimity about what heuristics is, or problem solving, good teaching, and so on, but I am convinced that in principle you are closer to my interpretation than you indicate in your manuscript, which I consider to be a mistake.'

J. DE LANGE

BETWEEN END AND BEGINNING

Mathematics Education for 12–16 Year Olds: 1987-2002

ABSTRACT. 2002 is marking the epoch as considered here. It is still within the interval Freudenthal once made predictions for, rather provocative predictions, I must confess. I quote a lecture from 1975: 'What to predict now? How mathematics education will look like in the year 2000? The answer is simple. There will be no mathematics education anymore in the year 2000. It will be gone. There will be no subject, called mathematics anymore, no mathematics lesson on the schedule, no textbook to teach from. Don't call me 'this devil's advocate', I àm the devil'. In the present contribution Freudenthal's predictions will be both confirmed and refuted. This means mathematics education is underway to getting integrated in other subjects to some extent, but on the other hand.... After an example of geometry from a new program, which is on blind spots captains of ships have to deal with while navigating, the arguments justifying the renewal of mathematics programs in secondary education are considered. Then the essentials of the olds will be characterised globally. However, what is left then, are the problems of implementation. The problems for the teachers evolve from the decisive differences between the old-fashioned rules and routines oriented programs and the new, problem oriented programs, containing a wealth of situations embedded in all kinds of interesting contexts. Differences in structure, in goals, in teaching, in tests and exams pas the review. Modern technology and its possibilities and promises are of my concern in the next chapter. Finally some conditions to be created by policy makers, are elaborated which to some extent would guarantee some sound development of mathematics education up to the year 2002. One of Freudenthal's predictions seems to come through: In the year 2000 pupils in general, secondary education will pick up more mathematics than ever before, or so it seems.

1. INTRODUCTION

A good example bears following, but examples, isolated and taken out of context, often give a distorted and idealized image of reality. Time and again they prompt questions such as: "Fine, but what comes before and what follows after?"

Those are very legitimate questions and it is for this reason that the example at hand is a more elaborate one, a unit of several lessons for grade seven. It constitutes part of the learning material developed by the Freudenthal Institute for the American Math in Context Project. (1)

It provides an opportunity to illustrate a week of mathematics education, whereby the ground work is laid for much which will later be made explicit. It also makes it possible to outline the objectives which the developers had in mind, the structure involved, the instruction that should accompany one and other, and especially also indicate a possible suitable assessment that might be developed for it.

Talking, theorizing and philosophizing about mathematics education is fine, but it is better still if these discussions can to some degree be assessed against

Educational Studies in Mathematics **25:** 137–160, 1993.

Fig. 1.

Fig. 2.

everyday practice. And practice often proves to be different than one imagines from behind one's desk hence that observations during this kind of developmental research are of such vital importance.

1.1. Ship Ahoy

(1) You are swimming in the canal and there is this big barge coming your way. It comes closer and closer. In the first photograph the captain cannot see you, you are still too far-away. The captain cannot see you in the second picture either. Explain why not.

(2) Of course how near you can get to the front of the boat and still be seen also depends on the shape of the boat.

Draw the line of vision for the captain of each of the boats below (the captain's position is marked with *).

Fig. 3.

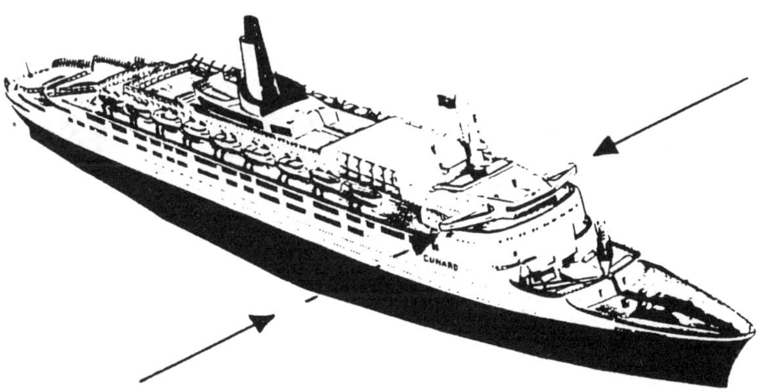

Fig. 4.

Measure the angle between the water surface and the line of vision in each of the situations. What does a small angle mean?

(3) Side-views of boats and the lines of vision do not tell you everything about what the captain can see

Between the two arrows you see the bridge of the Queen Elizabeth II. The captain and the first officer steer the ship from the bridge. Why do the captain and his crew have a good view on the water in front of their ship?

How does the shape of the area that the captain cannot see change when he walks from the extreme left to the extreme right of the bridge? (If you wish you draw your answer).

(4) Hydrofoils have underwater vanes which lift the boat out of the water at high speeds (figure 5).

Fig. 5.

Fig. 6.

Make side-views of the boat when it is in the water and when it hovers above the water and describe the influence the hovering has on the non visible area on the water (draw your own side-view and use your imagination).

In the first picture that boat is going very slowly, in the second it is going very fast.

Explain the difference in what you can see; make a side-view of the boat for both situations.

The barge in the pictures in figure 1 kept coming closer and the captain did not see you. In order to see the water better captains often follow a zigzag course (Figure 7).

Explain why the captain has a better chance of seeing, you this way.

1.2. The 'Blind Angle' (an Activity)

Bring a small plastic or wooden boat to school.

Compare the boats that are brought and predict which of the boats will have the biggest and which will have the smallest blind angle. The blind angle is that part of the water surface that the captain cannot see.

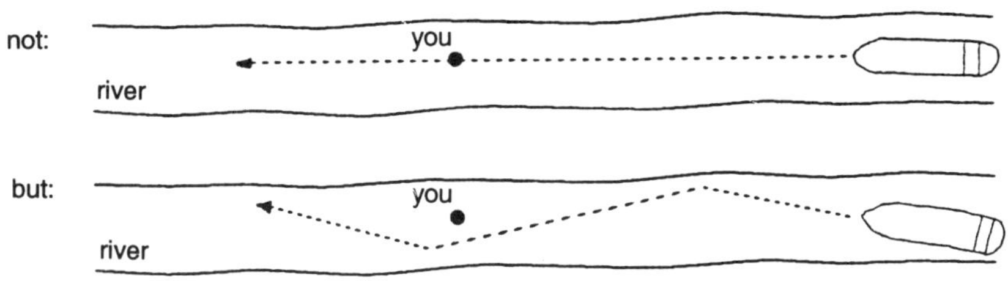

Fig. 7.

We are going to make the blind angle visible for the captain of your boat. What you need: your boat, a piece of cardboard, a few pieces of string and thumb tacks.

1. Place your boat on the cardboard, like in picture 1 (Figure 8). Make an estimate for the area (the water surface) that the captain cannot see.
2. Construct a line of vision straight ahead, like in picture 2.
3. Construct other lines of vision, like in picture 3.
4. We will now determine the ratio between the surface area of the blind angle and the surface area of the boat (from the bridge to the bow). Pictures 4 and 5 show you how you can do this. Now calculate the surface areas and the ratio for your own boat and compare it to your estimate in question 1.
5. Some tug boats do not tow, but push. These are pusher tugs (figure 9). You can make a model of a pusher tug with blocks. First try to estimate the shape of the blind angle. Now draw the blind angle. You can use real cubes and string, or use the work sheet.

1.3. Reflection

Freudenthal would have been delighted with this version of 'Ship Ahoy'. The grand old man and his grandson walked on many an occasion along the canal near his home. He recounts such an experience in 'Strolls with Bastiaan'. "Observations were made, one after the other. There was a great deal going on in his mind and also in Bastiaan's."

We imagined what a dialogue between them might have sounded like.

H.F.:	Do you see that ship over there?
B:	Yes.
H.F.:	Can the captain at the wheel see the man in the row-boat cross the canal? ....
B.:	How can I tell?
H.F.:	Well, look carefully, and think of the question I could ask you right here.
B.:	Well, if I were standing at the captain's side then..."
H.F.:	How can the man in the row-boat be sure?

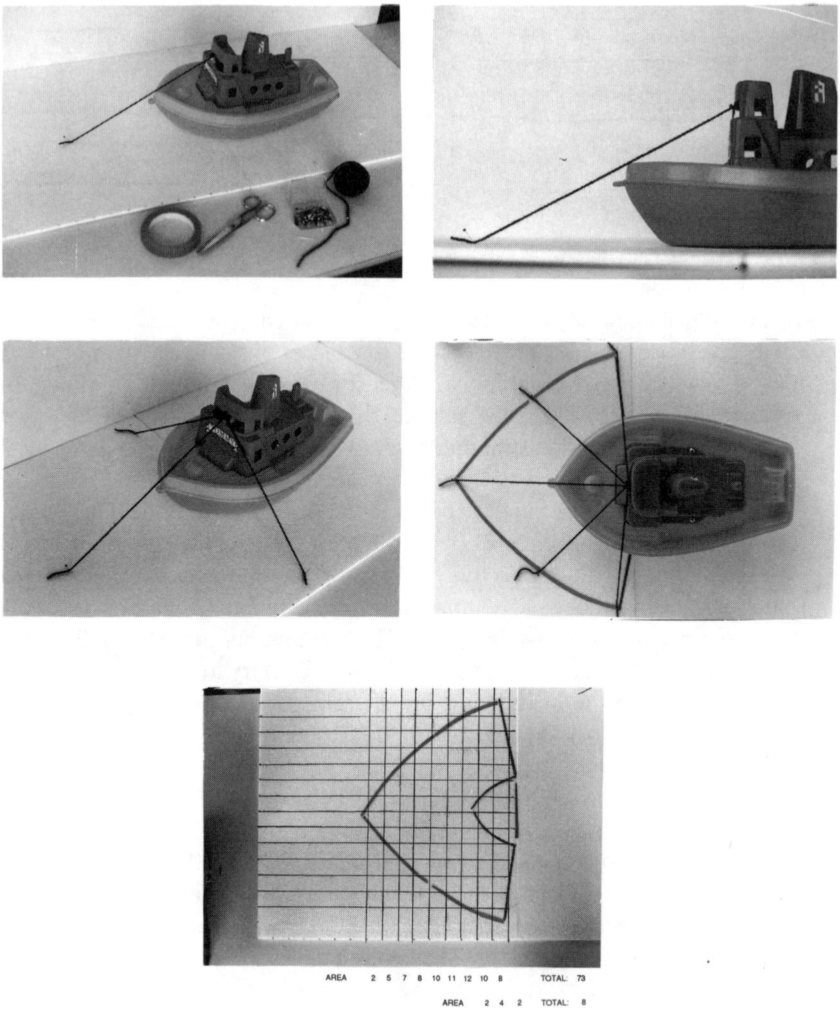

Fig. 8.

Of course this is mere speculation, but on the other hand...

Several years later Freudenthal published on the development of reflective thinking and the role of reflective thinking mediating between mathematical construction and proof (1979a,b).

'Ship Ahoy' is so demanding for pupils in respect to the mental activity of shifting, that is not mere speculation any longer to suggest its contribution towards the development of pupils' reflective thinking, or is it? Freudenthal would indeed have appreciated 'Ship Ahoy' and so would Bastiaan.

Further comments are superfluous, except perhaps drawing attention to the

Fig. 9.

suggested interaction in the thought experiment. The reader might expand on it by considering:

- the import of interaction in the classroom
- its possible influence on the development of meta cognitive abilities in pupils, and last but not least:
- how to reflect on mathematics education in general?

What we should start with in my view and what we should return to, not incidentally, but time and again, are genuine practical examples connected with paradigms of individual learning processes. That is what counts, that was, will and must be our source of inspiration. Let Freudenthal's spirit pervade the haunts of math education in this regard.

2. THE LAST GAP IN SECONDARY EDUCATION

In August 1985 a new curriculum was introduced in The Netherlands at an upper secondary level: Mathematics A. This curriculum was aimed at students who were preparing to study psychology, social and economic sciences at university. It was a curriculum that was considered by many as revolutionary because it broke away from traditions in mathematics instruction. The Hewet experiments started at two schools in August 1981, in the 11th grades. There experiences were to be continued in 12th grade and culminated in an exam in May of 1983. While the project at the two schools continued, they were joined by ten more schools in August of 1983. In May 1985 pupils at these 12 schools took the exam. These exams were only on the new Mathematics A curriculum. In August 1984, 40 more schools joined in the experiments, by this stage involving both the new A and B programs. According to plan the new programs were introduced at all schools in August 1985. A similar project started in 1987 with the purpose of producing a new curriculum for upper secondary, not university-bound students. An almost identical method of innovation was followed, although on an even smaller scale. The project started out at three schools and two years later the experiment was expanded to include 25 other schools. In the spring of 1990 the

Hewet experiments were declared a success by the government and consequently the two new curricula were introduced in August of 1990 (de Lange 1987). In our opinion, prematurely and involving risk for the immediate future.

The final gap in the introduction of the new mathematics curricula is the age group of 12 to 16. The Freudenthal Institute and Foundation for Curriculum Development (SLO) worked together during the period 1987–1992 to develop a new curriculum. The final report was submitted to the Department of Education in September of 1992. It is expected that the department will introduce the new curriculum in September 1993, starting with students aged 12.

3. ARGUMENTS

Although the new curricula for upper and lower secondary education each have their own specific characteristics, they derive their grounds for existence and justification to a large degree from the same facts.

3.1. Changing ideas about the Place and Function of Mathematics in Education

The primary objective of current mathematics education appears to be:

– preparing the student to function in society.

This is not a typically Dutch phenomenon. Also in other western countries this shift towards the forming of intelligent citizens and greater numeracy can clearly be discerned. Society demands qualities from its citizens that are closely related to the learning of mathematics: the analysing of processes, the interpretation of results, assessment of mathematically shaded presentations, the understanding of the physical world in which we live.

A second objective is then:

– preparing the student for future eduction.

This objective is one of all times, but the substance thereof is subject to major changes. Where previously to biologists a book on mathematics was first of all a math textbook, today it is much more the application that is the point of issue and the mathematics that is presented is applicable or useful for that field. User mathematics has made a very natural appearance on the scene, supported by the fact that there are an increasing number of areas of application for mathematics and that applications also appear to offer the student support.

The traditional objective of mathematics thus became the third objective:

– to introduce the student to mathematics as a subject area (or as a game or art).

The unfolding of mathematics as mathematics is increasingly being delayed until later grades, if it has any meaning at all. It is not simple to characterise the current content of the upper secondary program as 'genuine' mathematics.

Upon the introduction of social and user mathematics the idea of 'proving' also lost ground. Sadly it deserves mention that also this part of education had already lost much of its relevance anyhow. A reconsideration in this regard in relation to 'common sense' reasoning does not seem superfluous (Freudenthal, 1991).

Besides these changed objectives other arguments can be found which have led to changes.....

3.2. New Insights in Learning and Teaching

Already since the seventies foundations have been laid for an approach to mathematics education which has finally now found its way into secondary schools. Learning is no passive process of absorbing information and the compart mentalised storage in the mind. Learning is something that must be done. For example, students are able to construct mathematical concepts partly on their own, on the basis of 'real' problems. That gives opportunity for own strategies, own levels, interaction and reflection on the followed process. It is a bonus that this concept supported by cognitive psychologists also fits into the shifting of objectives (Romberg, 1991).

3.3. New Math

The Netherlands is known to respond swiftly to new developments in the field of mathematics. Discrete mathematics especially prevalent since the computer has already found a permanent place in classes in higher secondary school. Graphs and matrices and problems of optimising are important new areas in applicable mathematics which were quickly implemented in this country. Another new subject area, that of exploratory data analysis (still gaining in importance) has already made its mark in new curricula.

3.4. New Technology

An often employed alibi for innovation is the introduction of new technologies. The introduction and development of the computer has, as mentioned above, given new impulses to subject areas. Unfortunately the computer in the school has not yet rendered the expected positive learning effects by a long shot. The computer continues to be a marginal tool of which the influence remains rather limited at present, at least at most schools.

But the future is just around the corner: the pocket computer, which does the questions on final math exams in no time, has become inexpensive. The hand held computer with a 'derive' function, is available for somewhere around one thousand guilders. It is not without reason that the Ministry of Education has already been asked to revise the program for upper secondary mathematics because of the graphic calculator.

3.5. New Society

Not only has mathematics become more socialized, society itself is subject to swift changes. An information society demands accurate assessment and valuation of data. But also changes in the population spread in The Netherlands set demands for mathematics education.

The role of women in society has changed and will probably continue to do so. The allochtonous part of the population is considerable and our mathematics is not always per definition their mathematics, to say nothing about contexts.

The changes did not simply come out of the blue, therefore. What is surprising is the fact that innovation has occurred from end to beginning: first the upper grades of secondary school, then the rest. The basis comes at the end: this is only possible in mathematics education.

4. THE PROGRAM

The COW (Committee for the Development of Mathematics Education) was instructed to deliver an examination program for the various streams in secondary education. The programs at hand are based on a great variety of facts: the changing objectives of mathematics education, the insights of the extensive COW committee, the insights of the executive Team 12–16 and the experiments conducted by the team at a small number of trial schools, the experiences of teachers and students at the trial schools, the reactions of many, mainly teachers to publications by the COW and the Team 12–16, the unsolicited reactions of interested individuals, official responses from the Dutch Society of Mathematics Teachers to mention only some.

A number of central issues can be taken from the report:

– *introduction of the topic (advanced) arithmetic*

This will hardly be a surprise to most. On the one hand it guarantees a better follow up after elementary school and on the other arithmetic skills are maintained, something which seems a bitter necessity. Reactions from the field show that great importance is attached to this topic.

– *a shift within algebra*

This shift is expressly the consequence of the earlier mentioned shift in objectives: more user mathematics, more interpretation, dealing more flexibly with various representation forms, less isolated from other topics. This has consequences for other aspects of algebra: it is no longer so much about algebra itself, less about manipulations with symbols, less about techniques in the traditional sense.

The discussion about algebra has been fierce and has also led to an ongoing adjustment and validation of current choices. However, what we have are only recommendations and the future will have to show to what degree the expectations can also be realized. It seems evident that a great deal of research and development remains to be done in this area. Those in the field have already made it clear that adjustment of the program is appreciated but that there are doubts about the degree in which this direction can be effectuated.

– *innovation of geometry*

Geometry was not really flourishing in the existing program. But this is an international phenomenon. Euclidian geometry suffered greatly from the denouncement by the Bourbakists at the end of the fifties: 'Down with Euclid'. The introduction of 'modern mathematics', based on the theory of sets, did much to accomplish the same. But transformation geometry soon also appeared to be a rather empty set.

In the meantime a replacement of sorts has been developed which can be called the geometry of looking: the 'what' and 'how' of looking are interesting questions by which to explore and understand the surrounding world. It also involves a great deal of reasoning but of a totally different kind than in the traditional program. One problem is the symbiosis with the aspects from the traditional program which proved worth maintaining. Much progress has also been booked in this regard, in part due to the many positive critical reactions from those in the field. The result appears to be a challenging geometry program. And that very much suits developments elsewhere in mathematics eduction.

– *a new plan for statistics and graphs*

As mentioned, ever more use is being made of statistics and exploratory data analysis. Not only in subject literature, but also in the media: newspapers, magazines, television, and so on. It is inevitable therefore that attention be devoted to this facet of 'social' math at an early stage. Objective is also to instil a critical attitude in students so that they will no longer merely take any graphic representation by some expert as 'scientific' and therefore dependable.

Graphs and networks, as well as the corresponding tables (later called matrices) have already been introduced in the upper secondary education. They offer effective means of visualizing reality and at a basic level allow for discussion about model forming. There is also an excellent link to combinatory counting, since trees are in fact also graphs.

4.1. Reflection

From the reported changes in the new program for 12–16 year olds, I take the 'changes in algebra' as an example. What these changes aimed at was finding the essentials of algebraising.

In his book published posthumously, Freudenthal (1991, Chapter 1) deals with some examples from (pre)algebra. Paradigms play a decisive role there, within the context of generalising, performing heuristic activities, streamlining and short-cutting, and so on.

The most striking aspects of his description of examples, apart from the role paradigms play, is the room for construction they offer the pupils. Their input might be considerable, even in the development of algebraic language. The didactical sequences as suggested by Freudenthal are very much in agreement with the new proposals for algebra in the new program for 12–16 year olds.

5. PROBLEMS

It all sounds very wonderful and promising, but reality is of course considerable more obdurate than the attractive implementation models and theories about mathematics education would like us to believe. Experiences with the rather abrupt implementation of the upper secondary school program and the experiments which preceded it, the experiments in the framework of the COW, as well as recent experiences with the realistic materials in the United States give cause for critical reflection. All we can do is touch briefly on the problems.

5.1. Routines

One concern that is raised each time innovation is at issue is about routines and skills. Pupils no longer have to know unusual products, no more cosine rule. Manipulations with completing the squares are out as well, to say nothing about exercises with fractions and fractional equations.

And such concern is justified. No one can however say at this moment exactly what the required basic skills are. Almost everyone agrees that we can do with fewer but where to cut down and why is a point of serious debate. In the program for 12–16 years olds this is also clearly the case. Committees and development teams have held many a heated discussion about the techniques in algebra and geometry:

- gone is detailed study of quadratic functions; now these are nothing more than one kind of function among many;
- gone are the products of the sum and difference of two quantities, at least the elaborate, almost mystical approach and exercise of this branch of algebra;
- gone is completing the square which was a fine way to achieve insight in solving all kinds of quadratic problems.

Is that progress? Or is that flower power in Dutch mathematics education, twenty years after the fact.

Of course something has come in the place of the supposed impoverishment: there is more attention for concepts, for interpretation, for qualitative approaches, for dealing flexibly with various representations and there is more integration of algebra with other parts of the math curriculum. But honesty demands one to remark that the last word about this aspect of innovation has not yet been spoken, nor can it be.

5.2. Reflection

Freudenthal struggled his whole mathematics educational life with the contradicting interests of the retention of insight and the training of skills, or so it seemed. In his last book he observes the shift textbook authors have made from the extensive to the intensive training of skills which seems to solve the tension between the two aspects of learning mathematics as mentioned before.

I leave the question of Freudenthal and the training of skills here for what it is, and refer the reader to the editorial and Treffers' contribution in this volume.

5.3. *Structure*

Mathematics education was clear cut in the sense that it was always based on an obvious structure. First linear functions, then the second degree functions and sometimes even third degree functions. First equations, then inequalities. First a short introduction, then some examples and then many, very many problems. Tranquil and easy to oversee. And not only for the teachers.

A student who participated in the Hewet experiment that led to the new upper secondary program once complained that the old program was a great deal clearer: you simply did the last problem of the lot and if you could do that one, you could do them all. For these new type of problems you have to do them all, you can't skip one.

The old structure also made teaching relatively easy because of the points of support it contained. Not so any longer. Now the structure is hidden under much text and context or is formed by a story that is not mathematical. Sometimes there seems to be no structure whatsoever.

In short, for teachers and students there is a point of concern here to be taken seriously. The development of a concept, the build up of an abstraction thereof, the transfer to other applications, all of these processes must be recognizable for the teacher if the intended education is to be realized. And currently that often leaves something to be desired.

5.4. *Objectives*

To be able to realize objectives in eduction it is evident that these objectives must be clear cut. However, often objectives cannot be discerned from the materials for pupils, as experiments with context enriched mathematics education show time and again. In the traditional programs this was much clearer: solving quadratic equations is an objective that can easily be realized. What is more, in it lies one of the attractions of all the fuss with quadratic functions.

But that is much more difficult for much of the experimental material from the various projects. The series of lessons about boats is one such example. Attractive, those activities, but why do it: what are the objectives, where does it lead, where is the structure?

A complicating factor is the increasing attention for process objectives. Critical reasoning, communicating, reflecting, finding various strategies, development of a problem-solving attitude, all of this now belongs to the key objectives of mathematics education. But precisely how one must achieve these objectives and through which activities; that is often less than clear.

It can take several years before the teacher rediscovers these aspects in his own interpretation of the materials.

5.5. *Teaching*

From the foregoing it follows that teaching has not become any simpler. Less structure in materials, objectives which are more difficult to recognize, vague higher goals, but also the interaction which is considered essential to good eduction, is difficult to organize. A good combination of classroom teaching, work in groups, individual work and discussion is not something that comes naturally. It requires a different role on the part of the teacher: less teaching and more facilities.

Add to this the fact that dealing with the various strategies of pupils, evaluating the same, the different correct answers that are possible, can lead to uncertainty on the part of the teacher. And a teacher is not supposed to lack confidence.

And the contexts can also be the cause for uncertainties: do the teachers know enough about them, does the book give all the information, are there any tricky questions? Even worse when pupils come up with bright solutions which the teacher had not thought of. Some feel that their authority is damaged in consequence. Fortunately, it appears in many cases that the enthusiasm of the students and the teacher's growing ease with the material will overcome much of this problem.

5.6. Assessments and Examinations

One problem that was greatly underestimated initially was that of the testing and examining of realistic mathematics education. It is one thing to develop attractive materials, making suitable assessments is quite another.

Higher process skills, fine, but how does one assess them? Not with multiple choice tests, that soon became absolutely clear. But how then? Something like the example below?

> Last summer Bill weighed 107 pounds. He lost 4 pounds and then regained 11. How much does he weigh now?
>
> Which of the following number sentences can be used to solve the problem above?
>
> a $107 - (4 + 11) = A$
> b $(107 - 4) + 11 = A$
> c $(107 + 11) - 4 = A$
> d $-4 + 11 = 107 + A$
> e $(107 - 11) + 4 = A$

(item taken from the Illinois state test in the USA (1989).

The item is open and the relevant form could simply be the problem as it is worded in the opening sentence 'Last summer Bill weighed...' But not so. Here students must identify the solution to the problem with a sentence that has a mathematical slant that no one would dream of using to solve the problem. It must also be borne in mind that the objective here was to solve the problems with algebra.

Another problem that never seems to be a problem is the fact that pupils are always given precisely the information they need. A situation that seldom occurs in everyday practice however. An item from the Illinois State test illustrates a praiseworthy attempt at presenting a less trivial form of a relationship between the given information and the problem.

Kathy has bought 40 c worth of nuts. June has bought 8 ounces of nuts. Which girl bought the most nuts?

a June
b They both bought the same amount
c Kathy bought twice as much
d Kathy bought one ounce more
e You can't know

Unfortunately another multiple choice question, but the attempt is admirable. It is interesting to alter this question somewhat and to study the change in value:

Each of the following four answers is correct under certain conditions. Describe which conditions will make each of these answers correct:

Kathy has bought 40 c worth of nuts. June has bought 8 ounces of nuts. Which girl bought the most nuts?

a June
b They both bought the same amount
c Kathy bought twice as much
d Kathy bought one ounce more

This way we have gotten rid of the multiple choice and gained even more: the student must figure out under which conditions a given answer is the correct answer. Thus the objectives to be assessed are immediately taken to a higher level and the answers will afford the teacher or tester far greater insight into the level of the pupil. That the test is more difficult to grade is a price we would gladly pay.

6. A LOOK AHEAD

6.1. Computer

The introduction of the computer in education has not delivered on first promises. The euphoria of the eighties has been reduced the last decade to an enthusiasm that prevails only in a small circle. And mathematics stands out favourably compared to most other subjects.

Initially there was a case of hardware directed innovation: the computers were carried into the school, some software coupons were handed out and there was some attempt at promoting expertise. It was however unclear exactly how eight computers were supposed to improve ones math lessons on a continuous basis. And that still applies today.

The quality of the software and especially the user friendliness of that software often left something to be desired. There were major problems with regard to organization in the school, everyone had to do with the same few computers. And what about didactics? What should computer assisted education look like?

Moreover, publishers were (and are) not all that eager to invest substantial sums in software. The market is much too small and one had every reason to rejoice if a publisher proved willing to publish software products. In short, computers are

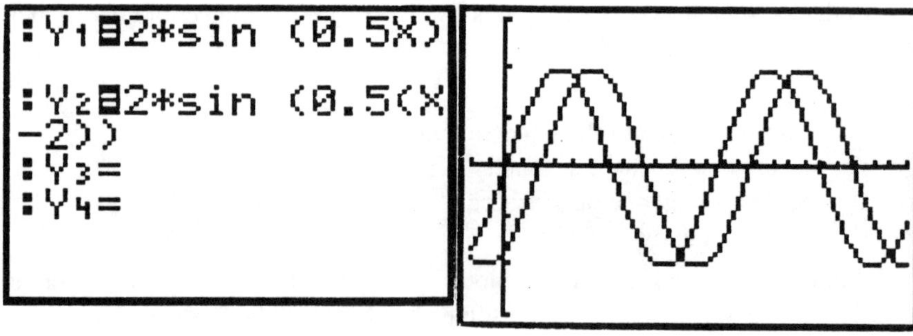

Fig. 10.

not the ultimate answer and it is the question if they ever will be. I think not, at least not the current and next generation of computers. According to the papers the Dutch deputy minister has now been persuaded that CDI will be the answer.

Mathematics in the year 2002? Technology will be an emphatic presence and it is up to us to ensure that justice is done to the realistic approach. Several developments lie ahead, some clearly predictable, others more speculative.

6.2. *Graphic Calculator*

The graphic calculator is capable of rather a lot: it draws graphs when a formula is typed in, it can zoom in and out, one can follow the graph, cut graphs, draw tangent lines, multiply matrices. Solving equations is no problem either.

The graphic pocket calculator is still in its infant stage; user friendliness must be improved upon, the screen is too small, the resolution not fine enough. The price tag is still rather high and the mathematical options too limited: one cannot differentiate functions (symbolically).

But it is obvious that this device will be capable of much more in a few years time, have an excellent resolution, work in colour, be able to differentiate and integrate and also solve differential equations. An example of the current possibilities.

The following graphs give a model-like image of the tidal curve of two Dutch coast resorts (Figure 10):

These graphs can be compared with the 'real' graphs of Flushing and Den Helder, as well as be analysed in regard to how they differ. In this 'real sense' students can be exposed to periodic functions at rather an early age.

6.3. *Reflection*

Structuring long-term learning strands at primary level was not Freudenthal's first concern, or so it seemed. Of course Freudenthal acted as the composer of didactical phenomenologies on primary subjects, but as Treffers states in his contribution, these were already somewhat labyrinthine at least from the point of view of designers of educational material. This, however, was not the case with secondary mathematics. On several occasions he pleaded for reconsideration of the approach to functions. In this sense he would have welcomed the graphic calculator. It would have been in support of his preference to start the concept of function with sine functions instead of linear functions. His long-term standpoint is reflected by the following:

> Sine functions is not a chapter to be rushed through in one breath as formerly was trigonometry. It is rather a learning strand, which by turns is loosened and tightened, according to what other intertwined learning strands admit or require; a learning strand the actual length of which has to be defined by each pupil individually. (Freudenthal e.a. 1989, p. 22)

It would be in concurrence with Freudenthal's views on the role of modern technology such as the graphic calculator that, after some handwork is done by the pupil, first for sine functions for instance, in an experiment with a turning wheel with the movement of a lighted point on its rim, the mathematical description of this movement, the choice of variables, devising a graphical representation, doing measurements and calculation based on Pythagoras for the sake of numerical specification etc. the graphic calculator would be introduced into the long-term teaching-learning process to proceed experiments with manipulation of the different parameters involved; cf. Streefland, (1990).

6.4. Interactive Video

CDI, or interactive CD, is an example of interactive video. The age of unattractive pictures on the computer, for instance in MS-DOS, are history. Video has taken its place. Although not all of the technical problems have been conquered it is evident that this medium is more interesting than the traditional computer with software.

An example that is almost technically feasible is a software program (Geometry Inventor) which consists of video recordings of such things as a walking elephant (figure 12). With the buttons at the upper left the elephant can walk, frame by frame, or in once from left to right and also backwards. We also get to see its paw prints. The movement of the elephant and the pattern of the prints render material for interesting discussion: for instance, the second picture shows three prints and for most adults and pupils it is very difficult to predict the next print (Figure 11).

The pattern also serves as a good point of departure to discuss periodicity. The zigzig/zagzag pattern is obvious, but which were made by the front paws and which by the back paws? Looking at the pictures one frame at a time gives the solution.

What about for other animals?

A trotting horse also has a zigzig/zagzag pattern (Figure 12)

But a galloping horse goes zigzag/zigzag (Figure 13).

But does the horse really walk the same way as the elephant or does it just seem that way?

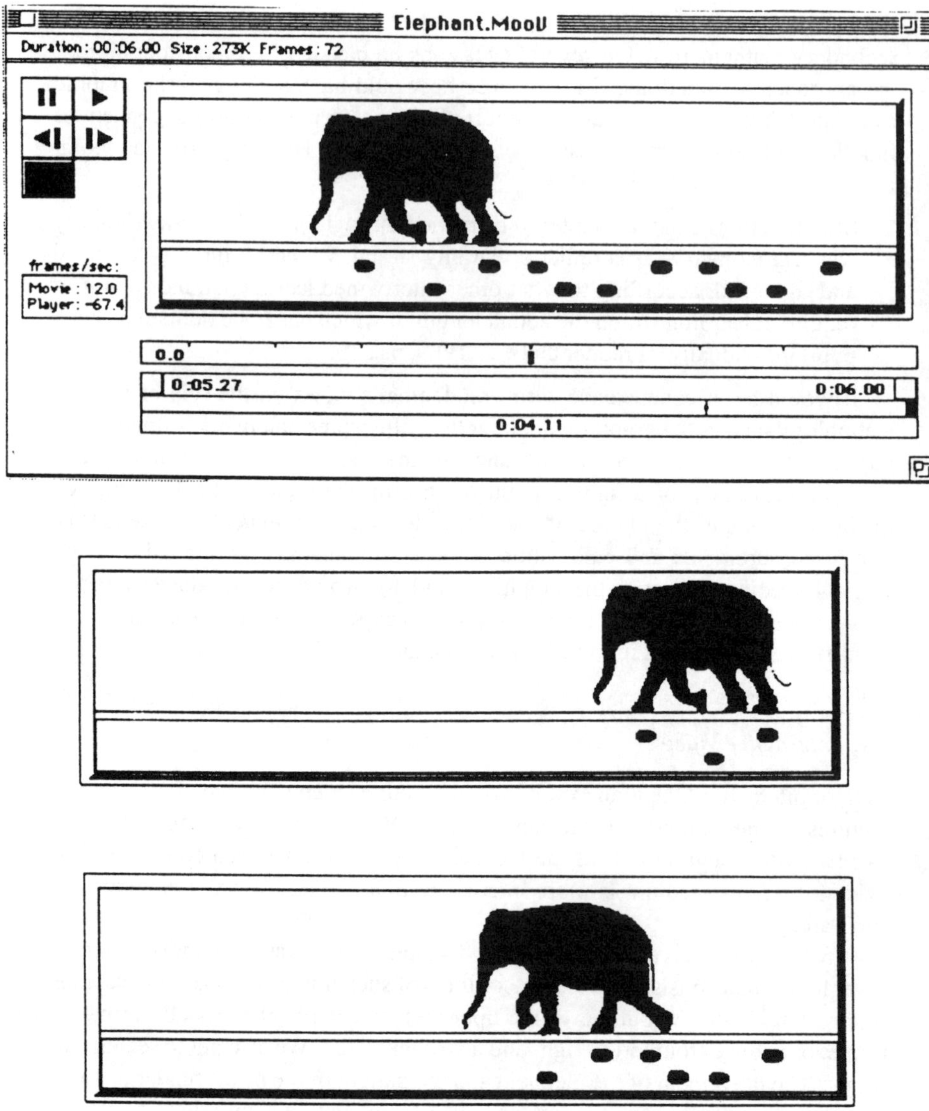
Elephant.MooV
Duration: 00:06.00 Size: 273K Frames: 72
frames/sec:
Movie: 12.0
Player: -67.4
0.0
0:05.27
0:06.00
0:04.11

Fig. 11.

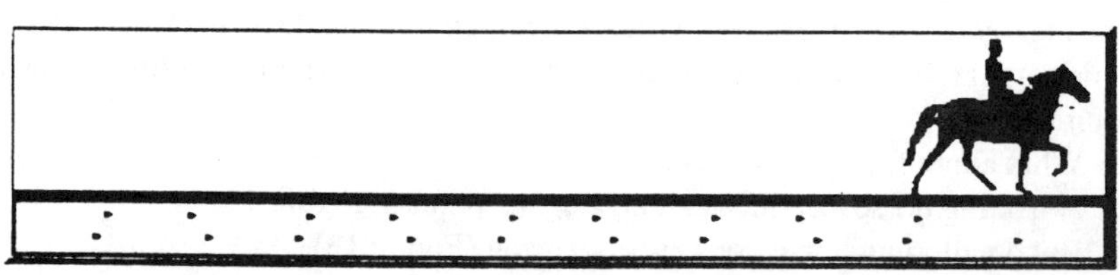

Fig. 12.

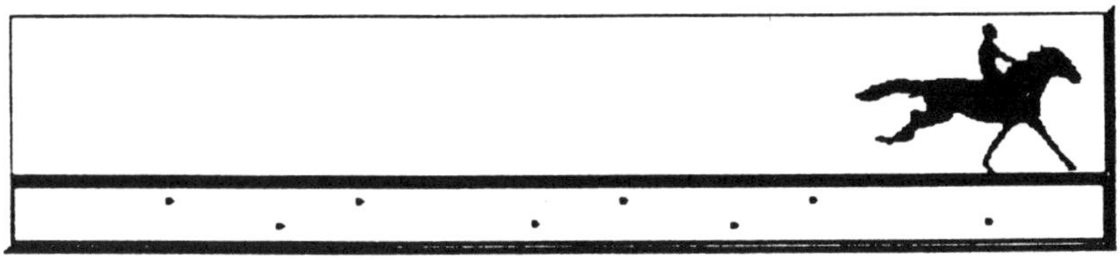

Fig. 13.

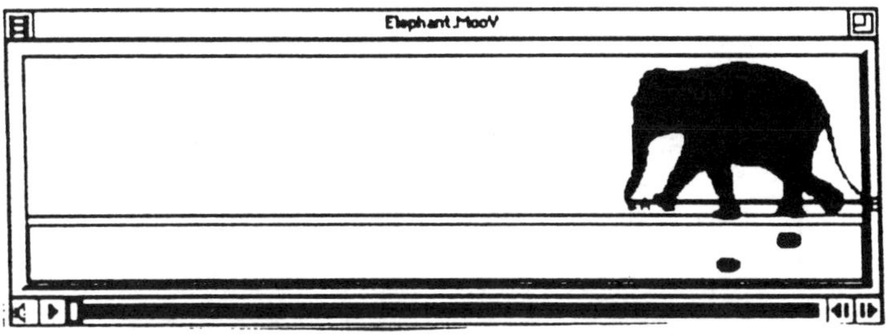

Fig. 14.

The film offers much material for reflection. But it really becomes interesting when we start manipulating the film: with the Geometry Inventor program we can (in the near future) draw point A in the trunk of the elephant and measure the distance to the starting line (AB) (Figure 14).

When the elephant starts to walk A stays 'in the trunk' and the line segment AB therefore becomes longer, while the clock starts running. At a random moment determined by us taking the pictures above as point of departure we can measure the distance AB and the time. In the near future one click will be sufficient: the table will be filled in automatically and the graph drawn (Figure 15).

This is a rather simple example: a simple time/distance graph, but it does illustrate the possibilities and the strength of interactive video/software programs.

Things become even more interesting when the horse graph is also drawn into the picture. But first let us make a prediction. And with colour we can do various things with the pattern of the prints: measure sizes and ratios of animals and humans. And then to imagine that we can input unlimited video and can let loose this software, which integrates the algebra and geometry, on it (Figure 15).

Mathematics research for primary school and secondary school students can in principle be further enriched as a result.

6.5. *Reflection*

A very brief annotation will suffice here. Freudenthal published a great deal on geometrical activities for young children, which means world- orienting activities varying from laying paving-stones to looking at the moon. Later on, as was explained elsewhere is this publication, he criticised the geometrical purity of his ideas and admitted that he would, if ever dealing with geometrical activities again,

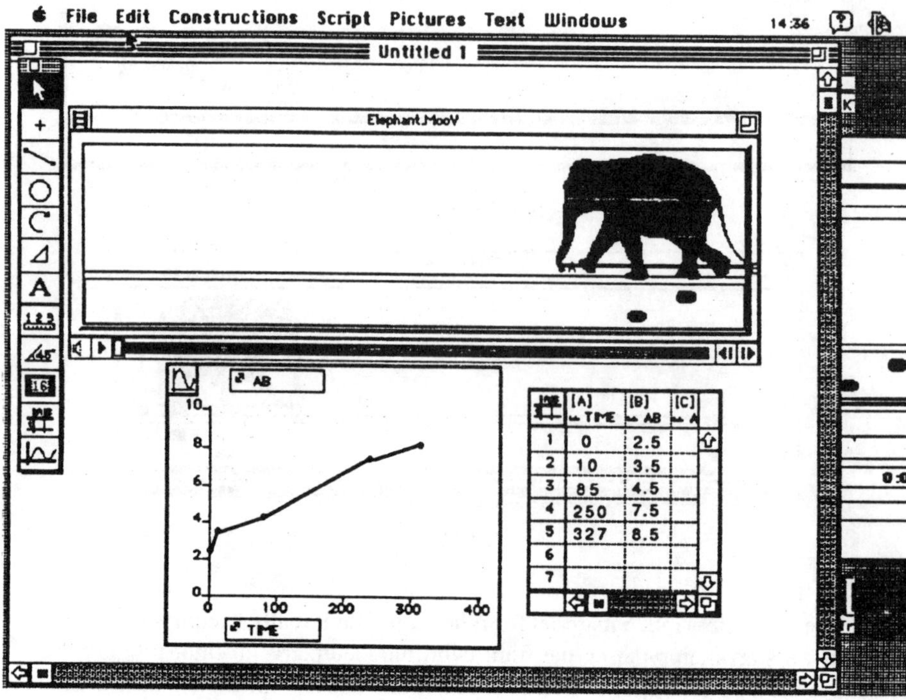

Fig. 15.

intertwine its learning strand with that for algebraic activities, not only for the sake of intertwinement as such, but in order to anticipate on future learning.

Well here it is, a posthumous fulfilment of his wishes, presented to him on a plate. He would have given his comments for improvement and at the same time have appreciated its possibilities for mathematics education.

7. CONDITIONS

A 'true revolution' is the qualification many give to the changes in Dutch mathematics education. And it does not look as if we are in a time of peace and quiet in the years ahead. The problems as outlined earlier will have to be solved, programs adjusted, technology be integrated, under condition that this leads to improvement of quality.

Is it as simple as that? The answer is an unqualified no. We mention a few points in this respect.

7.1. Evaluation and Examinations

It is astonishing to have to conclude that there has not been any kind of evaluation of the introduction of the new programs for upper secondary education. A brief

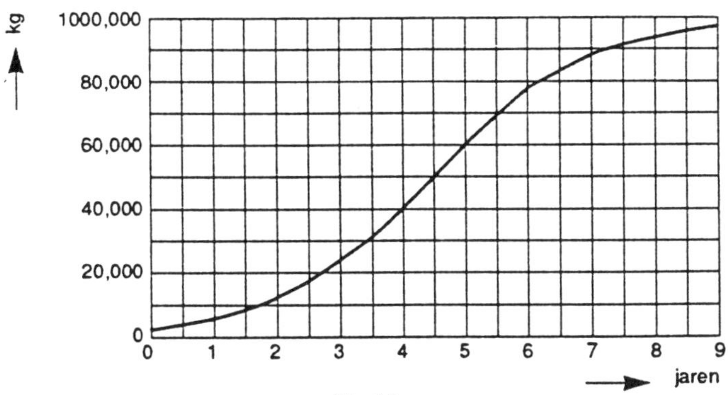

Fig. 16.

experimental phase, introduction by government order and that was it. Whether the new programs work, require adjustment, or whether better or different exams are necessary, all of these are issues that have been left unaddressed. A sorry state of affairs.

Very soon after the introduction of the exams the ministry and the commission which is responsible for the exams were approached about the fact that the exams did only in part assess the objectives which the curriculum was supposed to operationalise. Only once did we encounter an item of the following type:

> At a fish hatchery a number of fish are stocked in new ponds. If no fish are caught the fish stock will grow in the course of years. The graph below shows a model of the growth of the fish stock (Figure 16). 1. Draw the increase diagram for intervals of one year, starting with the interval 1–2. The owner of the hatchery will wait a number of years before catching. Thereafter he wants to catch the same amount of fish annually, as much as possible. The catch takes place each time at the end of the year. After each catch the fish-stock will again grow according to the graph above. 2. What would your advice to the owner of the hatchery be about:
>
> – the number of years he must wait after stocking the ponds
> – the size of the annual catch?
>
> Give an explanation for your recommendation that you think will convince the owner of the hatchery.

A truly open, especially qualitative problem which does not only test basic routines but also communicative skills. A question to which pupils have given some remarkable answers, like the pupil who took the humorous approach:

> I would wait four years for a catch of 20,000 k per year, boy, then we're business.

There were others who went as far as writing real letters:

> If you wait until the end of the fifth year you will have the biggest annual catch, namely 20,000 k of fish, and that is some catch! You can catch 20,000 k of fish each year without depleting the fish stock. If you were to wait one

year less you would have only 17,000 k left and almost one year later only 18,000 k of fish. So your yield will be biggest if you wait five years, so do not be impatient and wait (for your own good).

In all the years since 1987, the first national exam for upper secondary education, this was the only example we were able to find that made any attempt to test something out of the ordinary. Of course calculation and the drawing of answers are important and should be found on every exam, but if we are afraid to ask open questions on exams then we will not see our objectives upheld. And we can see evidence of that in current upper secondary exams.

Whence this fear for more open problems and process questions? Tradition in the sense that mathematics was precisely the subject that can be graded so objectively? Distrust of the expertise of teachers who are presumed unable to process the standards and scoring of open questions fairly?

7.2. Reflection

Freudenthal was very critical about formal testing techniques but he did observe some interesting changes taking place:

> My criticism of formal testing techniques with regard to research in mathematics education may appear to be outdated: There is an unmistakable trend away from it among educationalists with a strong mathematical background, who have good reasons to do so and are courageous enough to no longer be afraid of being reprimanded by methodologists for neglect of 'objective' methods. Open testing, interview, and observation of learning processes are gaining ground, and well designed intersubjectivity is given a chance to supersede ill conceived objectivity. Little attention, however has been paid to the methodology behind the new approach. J. de Lange's work is quite exceptional here (de Lange 1987). Anyway, there are promising perspectives on fresh approaches. Will they be successful in the long run? (Freudenthal, 1991, p. 155)

7.3. Development and Adjustment

The mistake in introducing the new upper secondary programs does not only lie in the virtual lack of evaluation but also in the fact that there is no mechanism whatsoever that provides in a continuous or discrete adjustment of the programs. It is evident for insiders that the upper secondary program requires adjustment in more than one sense. Perhaps the situation for introduction of the program for 12–16 years olds is somewhat more promising because it corresponds with the politically so important introduction of basic forming. There is an implementation plan that looks reasonably good and is already been executed; the guidance of teachers is getting more than minimal attention.

However it remains unclear how possible adjustment of the program is to take place, how signals from the field must be translated into action, and how to achieve that other objectives also necessitate different tests.

7.4. Training

There is no end to the complaints about teachers. Not in official documents of course, but all the more between the lines. The level is dropping, there is too little in-service training, there are no academically trained teachers any more, there is a lack of willingness on their part to innovate. Even the number of math students at the universities is related by some experts to the declining quality of teachers.

In this these critics are passing over a number of facts. The mathematics curriculum has changed drastically and has been accepted and carried out by the teachers with remarkable flexibility and a great deal of complaining. The Dutch teacher is seeing a record number of noncontent related, but structural innovations poured out over his head, from which only the politicians seem to have any gain. Many of these innovations do not make it in the end, but the unrest remains. There were the lyceum plans, compulsory math, modulation, basic forming and recently profile plans.

The Dutch teacher also teaches a record number of hours per week: 28 hours is unheard of in any civilized country, and that in excess of all the meetings the teacher is required to attend. The average Dutch teacher is not even physically capable of following in-service training. The conditions are deplorable: in-service preferably to be followed after school hours, at no extra pay and almost without any consequences for salary improvement.

Condition for the fulfilment of sound, responsible teaching:

- fewer hours, for example 21 as a target on the longer term
- more in-service training, in part during school hours
- reward for increasing skills, for instance as an extra periodical or a trip abroad to a seminar or such
- less structural organizational and better supervised substantial changes
- more involvement by teachers in the substance of innovation.

It sounds dramatic, but this merely outlines the situation that is common practice in some other countries.

8. CONCLUSION

Mathematics education in The Netherlands has an excellent reputation internationally and this is justified in part. But the innovations require a great deal of follow-up, guidance, evaluation and adjustment. These activities must be combined with the innovations, notably in the area of new technology and in the field of assessment of the given education. The outlined problems will be solved to an increasing degree in the years ahead, but that will require much continued developmental research. Teachers will have to be given the opportunity to return to their profession that status and content that makes them the experts they would like to be.

Filling in 'the remaining gap' suggests the end of a long period of development. Nothing is further from the truth. It is the beginning of a major challenge: to give content to ideas that have not yet fully matured, to find solutions for the outlined problems, to integrate new media where appropriate, to find good assessment methods and give new content to the subject that is called 'math teacher'.

The end heralds a new beginning.

NOTES

[1] The Math in Contexts project is a NSF founded project for the development of a new Middle School Curriculum. It is carried out by the University of Wisconsin at Madison, together with the Freudenthal institute at Utrecht University, the Netherlands.
The unit "Shady Business" was developed by Jan de Lange, Els Feijs and Martin van Reeuwijk and americanized by Mary Spence.

REFERENCES

Freudenthal, H.F.: 1975, 'Wandelingen met Bastiaan', *Pedomorfose* **25**, 51–64.

Freudenthal, H.F.: 1976, 'Wiskunde-Onderwijs anno 2000. Afscheidsrede IOWO 14 augustus 1976', *Euclides* **52**, 290–295.

Freudenthal, H.F.: 1979, 'Konstruieren, Reflektieren, Beweisen in phänomenologischer Sicht', *Schriftenreihe Didaktik der Mathematik. Klagenfurt* **2**, 183–200.

Freudenthal, H.F.: 1979, 'How does Reflective Thinking Develop?' In: *Proceedings of the Conference IGPME Warwick.*

Freudenthal, H.F. and S.L. Kemme: 1989, 'Sinusfuncties', *Nieuwe Wiskrant* **9**, 20–22.

Freudenthal, H.F.: 1992, *Revisiting Mathematics Education*, Kluwer Acad. Publ., Dordrecht.

Lange, J. de: 1987, *Mathematics, Insight and Meaning.* OW & OC, Utrecht University.

Romberg, T.A.: 1991, 'Epistomological Issues and Challenges to Assessment', *Studies in Mathematics Education*, ICMI.

Streefland, L.: 1990, 'Development Research and Tests – sine Functions as a Paradigm', in *Proceedings of the BISME-2*, August 23.–25., 1990 Bratislava.

Freudenthal Institute,
Utrecht University
Utrecht,
The Netherlands

HAND FREUDENTHAL: REVISTING MATHEMATICS EDUCATION

CHINA LECTURERS

Mathematics Education Library, Volume 9, 1991, Kluwer Academic Publishers, Dordrecht/Boston/London, 200 pp., 120,– Dfl., ISBN 0–7923–1299–6.

In 1987 Hans Freudenthal followed the invitation from the Universities of Beijing and Shanghai, and some Teacher Colleges to offer a series of lectures and seminars on mathematics education. Whoever visited China can imagine what an exhausting and stressing enterprise that is – in particular at the age of 82! Freudenthal, however, who never lost his curiosity and commitment, wanted to disseminate the ideas and the work of his institute. Well known and highly appreciated as a mathematician he was then listed as a mathematics educator who "tried to display and illustrate (his) view on mathematics education in all aspects." (p. xi). However, bound to their tradition and cultural experiences and lacking prerequisites and information, his audience had difficulties to understand his particular approach to mathematics education. Although celebrating and highly admiring Freudenthal, the Chinese mathematicians and educators could not follow him in full extent, and when I visited China two years later, they still strived for an understanding of his lectures and asked me to "explain Freudenthal". But contrarily to most of his colleagues he interpreted the failing understanding he had encountered as an obligation and challenge to revisit his ideas on mathematics education and search for a better and more comprehensive representation. The book resulting from this endeavour – the 'definitive' Freudenthal as Alan Bishop calls it in his preface – is an attempt at re-assembling the principal ideas regarding mathematics education he had developed during some 40 years and at giving them a new, clear and convincing presentation: "Resuming all old ideas, my aim is now comprehensiveness, not by compiling, but by selecting and streamlining, by including essentials, eliminating contradictions." (p. xi)

Setting up a frame, for the expectations of the reader the book starts with an 'apology and a warning': "The present book adds nothing but itself to work that I have published in the past in various places. I have renounced originality, I have refrained from offering the reader any new experiences, aspects, or ideas he might feel entitled to expect...

Views, even when supported by evidence, are susceptible to change. My present viewpoint is 'here and now', that of review rather than of overview." (p. xi)

The subtitle of 'China Lectures' is meant as an homage to my listeners who, by their critical curiosity, induced me to improve on clarity and concreteness." (p. xi)

Freudenthal dedicated this book to them who had not yet really understood him. A clear understanding is a prerequisite of a critical appraisal. Freudenthal knew that his work deserve acceptance or reputation on grounds of serious argument, but before all deserve discussion. We may regard this book as an ultimate offer to embark on discussion.

The 'China Lectures' contain three main chapters which are devoted to one major area of consideration, namely: "mathematics phenomenologically" (chapter one), the presentation of his "didactical principles" (chapter two), the outlining of the "landscape of mathematics education" (chapter three).

The major didactical principles which serve as guiding forces in Freudenthal's conception of mathematics education essentially rely on his perception of mathematics and of the approach to learning of mathematics. In the first chapter therefore, Freudenthal offers *his* answer to the question: What is mathematics? In Freudenthal's view, mathematical concepts, structures and ideas serve to organize the phenomena of the real worls as well as of the field of (developed) mathematics. The development of mathematics, the learning of mathematics as well as the creating of mathematics, starts from "common sense"; learning and developing mathematics are conceived as an ongoing process of mathematization on different levels and into different directions: basic activities within this process are local and global ordering, schematizing, formalizing and symbolizing, and – most important – deliberately reflecting and communicating these activities.

In chapter two, three main didactical principles characterize his didactical perspective: mathematics education as guided reinvention, as connected to reality, as focusing on learning processes with long-term orientation. As following from the view of mathematization displayer in chapter 1, teaching and learning have to start from rich contexts in which mathematics could be studied and experienced in use or on work: the applications of mathematics play a major role in this conception, in contrast to recent reform movements like the "New math" in which applications – for systematic reasons – were almost completely neglected and pushed aside. Learning mathematics, in his view, is necessarily based on exchange with others, and consequently social interactions became an indispensable precondition, and the interface between individual and collective experiences must be enforced by intensive communication and experimental work in heterogeneous groups.

In the last chapter Freudenthal discusses the "landscape" of theory, research and practice of mathematics education. Here he summarises his earlier critique of some features in the field of mathematics education, theory as well as research, and displays the rich developmental and practical work of his IOWO institute, later OW&OC working group at Utrecht, The Netherlands. Freudenthal's quintesential and most ambitious construction: mathematics education as "Mathematics for All and Everyone" emerged from the work with that institute. With the design of "Mathematics for All and Everyone" he connects the demands on the usefulness of school mathematics and the call for democratic participation in mathematical knowledge with the postulates of scientific standards of mathematics.

In the last paragraph of chapter 3, which is entirely devoted to this concept, Freudenthal makes it clear that he views it as a challenging task, not as ready – for – use programme, and presents some thoughts about its consequential meaning, as e.g. "Yet as much as I have refrained from discussing subject matter in the present book so little will I deal with the contents of Mathematics for All. Anyway, it will not be the same thing for each particular learner; there is a great deal of diversity not only for contents, but also or even more so, of breadth and depth of understanding... We teach classes, so why shouldn't we expect classes to learn? Life is cooperation. Why do examinations focus on individual performances and entirely disregard collective ones?... It will take much time and trouble before these words become deeds. A first step on this long path is to formulate, besides individual objectives, cooperative ones of Mathematics for All, and to describe specific tasks to be performed in cooperation with others." (p. 179/180)

It goes without saying that reading Freudenthal's book again is a pleasure, as it always was. As ever he is concrete and practical, even in his theoretical work and he is critical of detached theorizing as he is of footloose empirical research. There is formal thinking *and* an immediacy of both observation *and* judgement, philosophical perspective *and* an ironical, sometimes sarcastic, distance. There is a background of achievement in pure mathematics to support his engagement with education. The blend of all these characteristics gives his work a brilliant colour, and sometimes adds irritation. Buth they do more; they contribute a specific dimension to the developing science of mathematics education, a consciousness of the broad scope of its cultural context.

"Revisiting Mathematics Education" can be read as a synopsis of Freudenthal's previous books on mathematics education. It centers around some principles he summarized in his 1990 autobiographical sketch, as follows:

- "Mathematics as a human activity at various levels within the lived reality. This activity is described by the general term of
- Mathematising, with its various aspects of schematizing, formalising, algorithmising, structuring, axiomatising, and so on, and differentiated (after Treffers) into a horizontal and a vertical component.
- Didactical phenomenology as a way to describe mathematical ideas as being related to the phenomena that are to mathematise.
- Learning being understood as guided reinvention.
- Learning processes being articulated by levels (P. M. van Hiele's discovery, though modified by myself): the activity at one level becomes, by reflection, subject matter at the next higher level.
- Representing reality didactically by rich contexts.
- Mediating mental objects versus concept attainment.
- Mediating mathematical structures rather than the structure of mathematics.
- Retrospective and prospective learning.
- Fighting antididactical inversions.
- Training meaningfully enough to prevent sources of insight from being clogged by training.

- Developing mathematical attitudes.
- Didactising understood as a parallel to mathematising.
- Curriculum development understood as a learning process, in which thought experiments play an important part.
- Mathematics for All: teaching mixed ability groups."

Compared to Freudenthal's earlier principal books on mathematics education, in particular to the "Didactical Phenomenology of Mathematical Structures", the "China Lectures" are more concise and more balanced in their relation of concrete illustrative material and theoretical reflection. Thus the network of theoretical bonds becomes more visible, and the complexity and consistency of the theoretical foundations more striking. It might be fruitful to return to the overwhelming richness in concrete detail of the "Didactical Phenomenology of Mathematical Structures" after having studied the "China Lectures".

Yet the greater lucidity also pertains to some problems in Freudenthal's approach: There seems to be a contradiction between deliberately rejecting a systematic structure to the field and the requirement of organizing patterns at least for explanatory purposes. Freudenthal's heavy criticism of all kinds of systematic conceptualization of mathematics education and learning mathematics (e.g. the "New Math" approach or the behaviourist approach) is well substantiated: a closed, systematic and deductive approach would necessarily determine and suffocate the education process.

Freudenthal invented metaphors which – such as the concept of the "landscape" of mathematics education – appear very appropriate, as they may simply mean both, structure and its non-systematic character. Other concepts – like his relating mathematics to "common sense" – are less convincing, because they appear rather vague than complex. "Common sense" may refer to a nature-given nucleus of logic or to Platonic ideas, or it may mean common – i.e. social experiences in a highly mathematized society and some kind of social convention or contract. Further discrimination seems possible and necessary.

All these consideratons suggest to view Freudenthal, the person, as an inextricable part of his approach, not only as its source, but also as its focus. He no longer is. We look at the Freudenthal Institute. Their work will be an answer to these problems.

CHRISTINE KEITEL

Technische Universität Berlin
Fachbereich Mathematik MA 7-3
Straße des 17.Juni 136
D–10623 Berlin 12
Germany